Flying Snake

A Journal of
Cryptozoology, Folklore and Forteana

Volume 1 Issue 1   **April 2011**

I0023985

Was more like a Flying
Gecko than a Draco

membrane of skin
like a winged Girdle.

Watercolour sketch
of W'yang Gliding Lizard.

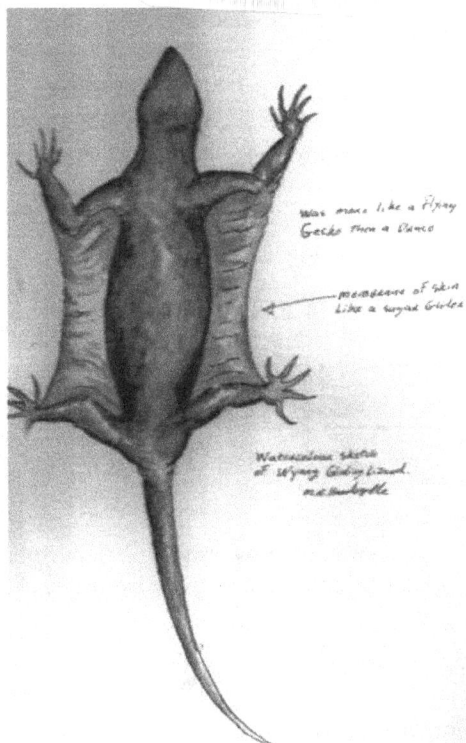

In This Issue: Pink-Tusked Elephants in Tang Dynasty China
•A Wild Cat in Dorset •Giant Centipedes of Hong Kong•
•An Odd-Coloured Badger in Derbyshire, UK • And More!

ABOUT FLYING SNAKE

Flying Snake can be bought from:

Flying Snake Press,
112 High St,
Macclesfield,
Cheshire,
SK11 7QQ
UK

Tel: 01625 869048

http://homepage.ntlworld.com/richmuirhead/cryptozoology/

My initial plan is to publish 3 issues of Flying Snake a year. Please feel free to contact me if you want to reproduce anything I have written. If you want to reproduce other authors' works, I will try and contact them on your behalf and get back to you. The opinions of authors other than myself do not necessarily reflect my own.

Back issues available on request.

I hope to publish PDF versions of Flying Snake as soon as I get a chance to learn how to use the technology.

PAYMENT

Subscriptions: £ 3 per issue, £ 9 per annum.

Payment for however many issues you (and your friendly neighbourhood, reading, flying, snake) would like to purchase can be made by means of PayPal on my web site (See url above).

Cheques and postal orders from within the United Kingdom should be made out to Richard Muirhead, please, and *not* Flying Snake or Flying Snake Press.

Cheques will not be accepted from outside of the U.K at this point in time.

CREDITS

Flying Snake Title/logo: William Frederiksen

Flying Snake image on front cover: Janet Bord and the *Fortean Picture Library.* Flying lizard illustration on front cover,Mike Hardcastle.

The Collected Flying Snake

Volume One

Typeset by Jonathan Downes,
Cover and Layout by SPiderKaT for CFZ Communications
Using Microsoft Word 2000, Microsoft Publisher 2000, Adobe Photoshop CS.

First published in Great Britain by CFZ Press

CFZ Press
Myrtle Cottage
Woolsery
Bideford
North Devon
EX39 5QR

ISBN: 978-1-909488-58-8

Foreword

I have known Richard Muirhead longer than I have known anybody else to whom I'm not related.

We first met in Hong Kong, in 1970, when I was recovering from major surgery on my legs, and – although we lost touch for some years – we made contact again in the early 1990s, and – to my great pleasure – I found that the grown-up Richard shared many of the same interests as me, particularly as regards cryptozoology and allied disciplines. So, when Richard asked me to write the foreword to this, the first volume of collected editions of his magazine, *Flying Snake*, I was glad to accept the challenge.

As anybody who knows me will attest, I have always considered the late P. G. Wodehouse to be one of the greatest 20th century exponents of the English language. As well as a peerless comic writer, he was one of the greatest wordsmiths of his (and many other) generations. One of my favourite passages of his, describes the museum at Blandings Castle in rural Shropshire:

> "In the museum at Blandings Castle you could find every manner of valuable and valueless curio. There was no central motive; [....] Side by side with a Gutenberg Bible for which rival collectors would have bidden without a limit, you would come on a bullet from the field of Waterloo, one of a consignment of ten thousand shipped there for the use of tourists by a Birmingham firm. Each was equally attractive to its owner."

Lord Emsworth's museum is just the sort of place that I like to visit, and appears to be a fine example of what the well-bred Victorian gentlemen would have called a 'Cabinet of Curiosities'. And so is this book.

Richard Muirhead has a remarkably enquiring mind, and he is one of those lucky people who – despite all the pressures of our decadent age – remains very much a generalist.

Therefore, his magazine, *Flying Snake*, contains a fascinatingly disparate collection of articles on various cryptozoological, Fortean, and phenomenological subjects; all filtered through his own inimitable viewpoint, and it is both a pleasure and a privilege for us to be re-printing them in book form. Richard's eccentric interests are mirrored in his equally eccentric typesetting and layout skills, and we have kept them just as they are so as not to compromise the emotional integrity of this unconventional and often whimsical publication.

I hope you enjoy this as much as I have.
Come on in, the water's lovely!

Jon Downes
Woolsery, North Devon
March 2019

CONTENTS

DR DEVO'S DIARY

"For I pray God for the introduction of new creatures into this island. Fo
I pray God for the ostriches of Salisbury Plain, the beavers of the Medwa
and silver fish of Thames." Christopher Smart, `Rejoice in the Lamb`.
Poet, naturalist, lunatic (1722-1771)

Hello and welcome to Flying Snake issue 1. In case you are unfamiliar with my alter ego `Dr Devo`, he only exists as a by-product of my relentless listening experience of Devo since about 1982. The nearest thing they have come to being exponents of cryptozoology or Fortean zoology is the song `Pink Pussycat` – as far as I know! But hey, if a green kitten can turn up in Denmark in 1995 then I guess anything is possible. In the book review section I look at the weird anthropology of `In The Beginning Was the End` by the equally weirdly named Oscar Kiss Maerth. This book title featured in Devo's song `Gates of Steel`. I also review the 3 volume `Moa Sightings` by Bruce Spittle.

Devo are basically a concept band, de-evolution being the belief that humans are de-evolving into apes, if not literally then morally. However, I digress massively, this is a magazine about cryptozoology, Forteana and folklore, not de-evolution! It is also deliberately "archive" (which I designate as pre-1950) orientated because I am interested in cryptozoology before "cryptozoology" was coined as a word by Ivan T. Sanderson in the early 1940s or Bernard Heuvelmans in the 1950s depending on your perspective. I am sure that this "antique cryptozoology" will throw up some classics; in fact it already has. Indeed my first foray into writing about cryptozoology featured an event which may have predated 1942 by many years, though the Namibian Flying Snake of 1942 at Kirris West was the incident I was concerned with in the mid 1990s, (see CFZ Yearbook 1996). Hence `Flying Snake` as the title of this magazine.

So, without further ado, read on! Comments, letters and criticisms to flyingsnakepress@hotmail.co.uk

Issue 1 of Flying Snake is dedicated to my mother Sheila Muirhead . Thanks to everyone involved in contributing and helping produce Issue 1.Thanks to Janet Bord of the Fortean Picture Library for permission to reproduce the image of the flying snake on the front cover.Thanks to Mike Hallowell, William Frederiksen, Helena Roberts and Nic Moran for help with production.

The Curious Creature

By Chad Arment

The following story was given to me as a transcribed report from the original newspaper article by veteran researcher Gary Mangiacopra. It is a good example of the sometimes strange nature of old newspaper stories.

"Nobody Knows Its Name—a curious creature found in the mountains of New Mexico

"Arthur Jilson and Al Quinn, of Las Vegas, N.M., have purchased the curious quadruped recently captured in the mountains between Albuquerque and San Pedro, and it is now on exhibition there. The animal was found in company with another of the same kind, but larger, which was supposed to be the mother of the one caught, but which doubtless was a male of the same species, as the captured one gave unmistakable evidence of being several years old.

"The pair were vigorously pursued and the larger one succeeded in making it's escape. the creature did not thrive in captivity and soon died. The body was vent to P. J. Martin, the Las Vegas taxidermist, who has mounted it in a very lifelike way, the position being that often assumed by a cow when lying down, the two hind legs under the body, the right fore leg doubled under the left fore leg extending to the front at full length, the head and neck lifted in the air and the ears bent forward as if listening.

"The animal, as described by the *Optic.* is something like sixteen inches high by twenty-four inches long, and has the appearance of being an inch or two higher in the hips than in the withers. Its hide is entirely hairless, and while the general color is a kind of purplish black, most of the belly, the front part of one side, one fore leg and a blotch on the other side are of a delicate fawn color, mottled with black spots.

"The lower portion of all the legs, from the knee down, is yellow, and the upper part of each leg, except the one already mentioned, is black. The tail is about three inches long and is quite small; the body has the general appearance of the cow; but is rather more slender and elongated: the legs are like a deer's with cleft hoofs; the head, neck, eyes and ears are those of a hairless Mexican dog, though somewhat larger, and with the additional exception that in the upper part of the forehead are two small, backward-curving horns, looking like those of a young goat.

"Altogether, the creature is very peculiar and no territorial naturalist has been able to decide whether it is a freak of nature, a cross between several animals or whether it belongs to some heretofore unknown species.

When open for mounting its teeth were found ground down pretty evenly with the gums, while the stomach was filled with grass. The animal will be sent to the world's fair with the exhibit from this county. Chicago *Herald.* (From the New Haven, Connecticut, *Evening Register.* September 6. 1892.)"

Researchers need to be very careful about accepting such accounts at face value. First, we know that"filler" articles were sometimes created when there were empty columns. Nature faker or wonder stories were not uncommon, as they entertained the reader with ideas of unknown phenomena beyond the recognised boundaries of civilization.
Even detailed measurements given do not necessarily argue for veracity, as some journalists who contributed tall tales recognised the value of creative details.

Still, an argument that a story could be a hoax is not an argument that a story must be a hoax, at least not without corroborating evidence. One strong argument against this being an impetuous newsman's fabrication is that at least one of the Las Vegas, NM, men mentioned, Arthur Jilson. was noted in a number of local news items during the 1890s as a sportsman (both hunting and horses).

Another possibility is that the animal was created for exhibition purposes.

Without knowing more about the backstory (and a general search hasn't come up with any further details on the specimen), it is difficult to say for certain that the animal was not some sort of taxidermy hoax. We know that exhibitors would sometimes create elaborate stories about where and how an animal was found, when the actual story (and actual animal) were much more mundane. If we start from the assumption that this could in fact have been a real animal, and the story was also reasonably accurate, it sounds fairly goatlike rather than deerlike. There is a (rare) hairless mutation that can crop up in domestic goats. The described size, however, is rather odd, and a small goat kid wouldn't have distinctively curved horns. (Suggesting a hairless pygmy goat seems an unnecessary complication, particularly as pygmy goats are an African breed unknown in North America at the time).

So. 1 really can't offer a likely explanation from known species. I've also not run across any similar reports, so this may just be another dead end in historical cryptozoological research. unless further details can be uncovered. The story does bring up two things to keep in mind regarding cryptozoology.

First, this is one of the few mystery ungulates in North America. Other cryptids tend to get better coverage, but there are some interesting cases of grazing mystery animals.

Second, there are still many more mystery animal sightings yet to uncover in historical archives. There arc certainly many more newspapers yet to be made accessible. Patterns in historical newspapers are of interest to cryptozoology whether they reflect folkloric motifs or real encounters with unknown species.

Are There Unknown Species of Flying Lizards Living in Australia?

By Mike Hardcastle and Richard Muirhead

In November 2010 I came into contact with Mike Hardcastle of Wyong, New South Wales who passed on his father's notes concerning the latter's sighting of (along side his brother) a lizard(see front cover image by Mark Hardcastle) which was neither the native frilled lizard, nor Draco volans of Indonesia and Malaysia. In this report we summarize this 1952 sighting and other anomalous flying lizard sightings elsewhere in Australia.

Wikipedia has this to say about Draco volans:

> *Draco volans, or the Flying Dragon, is a member of the genus of gliding lizards Draco. It can spread out folds of skin attached to its movable ribs to form "wings" that it uses to glide from tree to tree over distances upwards of 8 meters (25 feet); however, like all modern reptiles, it lacks the ability to sustain powered flight, and is capable only of gliding. Its wings are brightly colored with orange, red and blue spots and stripes, and they provide camouflage when folded. The flying dragon can reach lengths as long as 19 - 23 cm. It feeds on arboreal ants and termites (1)*

Here are the hand-written notes of Mr Richard Hardcastle dating from 1990 and late 2010. See also Fig 2 below.

Vicinity – Wyong district – Warnervale.

Time – Late 1940s

Information – From district timber cutters.

"Story was that one was killed in a tree that was felled. It was put on display in a shop window in Wyong for a few days, till it began to decay, then presumably thrown away. One description was about 10 inches long, with a flap of skin from fore to hind legs. Said to live along the creeks and glide from tree to tree like a sugar glider .Said to have been seen just north of Warnervale near the northern main railway line. Another report just north of Warnervale seen gliding from one tree to another size about 3 ft (?) colour brown. The creek in question was Wallarah Creek which runs from, or into Tuggerah Lakes from what was then bush and under the Pacific Highway near Charmhaven.

Sources of information, timber cutters (then) George Watling of Wyong and Merv Nicholson of Kanwal (now in their 80s if alive)

Olive green colour with a membrane of skin from front to hind legs. Skimmed from tree to tree .Found on Wallarah Creek…..(Wyong, Tuggerah Lakes area) Found by timber cutters in the 1950s. Was displayed in a shop window in Wyong for a few days Found between Warnervale and Wyee on the lake side of the northern railway line." (2)

Fig 1 Richard Hardcastle with the skin of a large, grey kangaroo.

Reproduced with permission of Richard Hardcastle

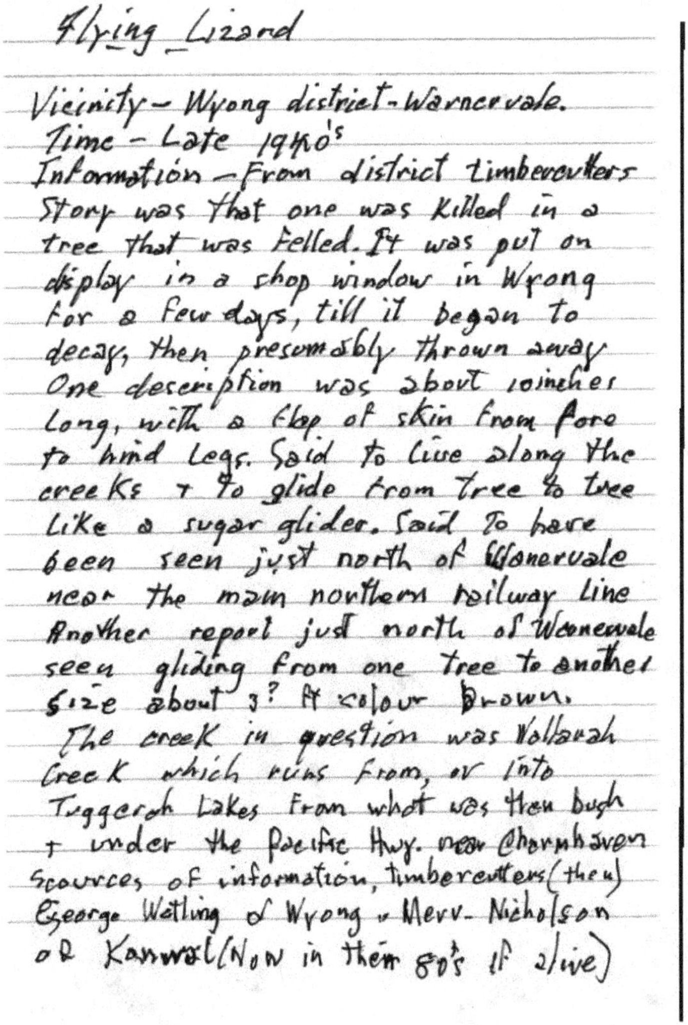

Flying Lizard

Vicinity — Wyong district - Warnervale.
Time — Late 1940's
Information — From district timbercutters
Story was that one was killed in a
tree that was felled. It was put on
display in a shop window in Wyong
for a few days, till it began to
decay, then presumably thrown away
One description was about 10 inches
long, with a flap of skin from fore
to hind legs. Said to live along the
creeks + to glide from tree to tree
like a sugar glider. Said to have
been seen just north of Warnervale
near the main northern railway line
Another report just north of Warnervale
seen gliding from one tree to another
size about 3? A colour brown.
The creek in question was Wallarah
Creek which runs from, or into
Tuggerah Lakes from what was then bush
+ under the Pacific Hwy. near Charmhaven
Sources of information, timbercutters (then)
George Watling of Wyong + Merv- Nicholson
of Kanwal (Now in their 80's if alive)

Fig 2 The handwritten notes of Mr. Richard Hardcastle

There is another account of this flying lizard in Rex Gilroy's `Out of The Dreamtime`:

A strange half - bird, half- reptile creature was known to the aborigines of the Murray River country. A large creature, it was said in one myth, to have entered a village in search of its baby which the child of one of the tribespeople had stolen. In the course of its search it killed two men. The myth of the half-bird, half-reptile creature was given publicity in a 1927 newspaper article, which even gave a supposed sighting of the beast by a European. The nearest creature this `bird` can be compared to is the Pterodactyl......In 1952 a Mr R.Hardcastle was cutting timber for a living at Warnervale near Wyong, on the New South Wales Central Coast. Thirty years later, in 1982, he related to me the following story:

" Some of the timber men I knew told me about " a flying lizard". They said they had seen them up in the forest country hereabouts over the year……[hereafter the account is very similar to the hand written notes quoted from above-R] Many of these flying lizards inhabited the tall trees near the creeks. My brother phoned reptile people at the Australian Museum in Sydney but they said there was no such species.!"(3)

The images below (Figs 3 and 4) from Google Earth show Wyong and Wallarah Creek where flying lizards have been spotted.

Fig 3 Google Earth image of Wyong area

There are reports of Draco volans-like lizards in Queensland and islands in the Torres Straight. For example, in the Pennsylvania Bristol Bucks County Gazette of June 20th 1895:

Flying Snakes of the Antipodes The race of flying dragons, which spread such dismay and terror in olden times, is not yet entirely extinct, if we are to believe the utterances in a late Queensland paper. The Queensland Mercury says: " James Bass of the sandy flat lying beyond the Blue hill, near the head waters of Carns' creek, has brought another specimen flying serpent to this office. It is somewhat smaller than the one exhibited by him at Gulley last year and

Fig 4 Google Earth Image of Wallarah Creek

larger than the one which he presented us on Christmas day. Like the other two, it has four three jointed legs, each 7 inches long. Between these legs, which are situated two on each side of the body, is a leathery membrane, much resembling a bat`s wing. Mr Bass declares that he has often seen them flying across Carns' creek at places where it is 60 feet wide." St. Louis Republic. (4)

On April 11th 1935 The Queenslander reported:

Flying Lizards by "WAYFARER"

Australia is rich in lizards,several hundreds of species being found throughout the Commonwealth,but it is not generally known that Australia has a flying lizard. This queer creature is not found on the mainland,but on certain of the islands in Torres Straits. The reptile is fairly plentiful on Moa Island, which is situated about 80 miles north-west of Cape York. The lizard is of a slaty colour, and it is remarkably like the common goanna in appearance; the differerence, however,is that it has "wings."These "wings" are simply pieces of strong but transparent skin stretched between the fore and hind legs. These "wings" do not enable the lizard to actually fly, but by means of them it is able to volpane to the ground from trees, after the manner of the flying squirrels. Because of this habit the lizard is greatly feared by the natives, for their have been dozens of cases where the lizards have accidentally struck natives when gliding to the ground. The reptiles have sharp claws and many of the natives so struck have been badly cut and scratched by the lizard`s claws.- "WAYFARER."(5)

On 16th September 1939, The Courier-Mail (Brisbane) reported:

"Flying Dragon" Sought

On a five week tour of investigatory and collection among islands off the north Arnhem Land coast, the Melbourne naturalist, Mr Charles Barrett, who for many years edited nature publications for the Melbourne Herald and Sun,left Darwin to-day in the Methodist mission`s motor vessel Larrpan.

Accompanied by his wife,also a naturalist,he will make a 120 mile trip,in the course of which he will visit the remote Weesel Islands. They will be the first naturalists ever to set foot on them.

Mr Barrett will endeavour to obtain a specimen of the flying dragon, a volplaning lizard believed to inhabit islands off the coast. (6)

Dale Drinnon has commented:

I mentioned Flying geckos to you last week in connection to some New World reports;however,I have heard independent mention of Flying lizards in Northern Australia(that spiky part that goes up towards New Guinea). If they are in novel areas they could be cryptids, more possibility of Flying geckos from Indonesia being released in Australia than in New World. (7)

The Argus, a Melbourne paper of July 25th 1911 reported a flying lizard type animal, though with Ropen-like characteristics from Warracknabeal, Victoria;

IN THE OPEN AIR

FLYING LIZARDS

A Warracknabeal boy sends a paragraph from the local "Herald" and asks if flying lizards are known anywhere. "Mr. T.W.Bochim on Sunday saw what he believed to have been a flying lizard. It was about a yard in length, of a greyish colour, whose head was very large. The strange visitor flew about 8ft above the ground.It crossed the yard of Mr Sims's residence, flew across Mr Bochim's property, and was lost sight of in Mr Taylor's yard. There is no definite knowledge as to whether flying lizards have previously been seen in the neighbourhood, and some information on the point would be interesting."

I can't suggest what this strange creature may be. There are no flying lizards in Australia. I have seen them in Java. They are beautifully coloured, but the largest not more than 8in or 10in in length, and live in the tree-tops. (8)

Two weeks later the same paper reported the following:

IN THE OPEN AIR

FLYING LIZARDS

Concerning the mystery of a flying lizard or some other strange creature at Warracknabeal, A.W Smith writes from that town "Some time before Mr Boehm's note on this matter, when coming home from work, one bright moonlit night , I was attracted by a rustling sound in a yard close by. On looking over I saw some dark but not clearly distinguishable object scaling a palling fence opposite. It went up quickly, with a distinct scratching noise on the pailings, that sound being quickly followed on the other side by a rapid whirr of apparently short wings, such a noise as a quail makes on rising. I mentioned the matter to several friends next day,but no one could account for it". (9)

Dr K.P.N.Shuker has noted a report of a volpaning frilled lizard from Queensland ;

Volpaning Frilled Lizard

Not long after the publication of this book's original edition, I learnt of an additional, equally controversial glider – none other than Australia's famous frilled lizard Chlamydosarus kingii. As already described and depicted in Chapter 2 of this present book, it is renonwned for the extraordinary crenated frill around its head and neck, which it expands if threatened, in order to startle and ward off any would-be attacker. However, some writers contend that it also utilizes its frill for volplaning, enabling it to glide from tree to tree.

In his book Bunyips and Billabongs (1933), Australians scientist Dr Charles Fenner included a remarkable statement made by Queensland naturalist Mrs Adam Black concerning this distinctive reptile:

...... a pair lived outside our garden fence for years. They would run up a tree if one approached, and I've often seen my husband put his hand round the tree (they always climb up the opposite side to where you are) and catch one's tail; he would then hold it and go round and stroke the lizard's back and frill. If really alarmed when up a tree they extend their gaily-coloured frill and glide down to root of another tree.

Sadly, Black gave no description of the volpaning itself. I can only assume that if it does occur, the frill must act like a parachute, opening out, thence enabling the lizard to drift passively downwards.

Summing up, Fenner stated:

I believe that Mrs Black and other observers have produced convincing evidence that we have an Australian "flying lizard."It is to be hoped that some zoologist will take steps to observe these volplane flights of Chlamydosaurus.

Unfortunately, this does not seem to have happened, so for now, this intriguing subject is very much up in the air. Whether the same can be said can be said of the frilled lizard itself, therefore, remains to be seen – literally! (10)

REFERENCES

1.Wikipedia Draco volans http://en.wikipedia.org/wiki/Draco_volans

2. Handwritten notes of Mr R.Hardcastle

3. R.Gilroy Out of the Dream Time-The Search for Australia`s Unknown Animals   (Katoomba: URU Publications,2006) p.294

4. Pennsylvania Bristol Bucks County Gazette June 20th 1895

5.The Queenslander April 11th 1935

6.The Courier-Mail September 16th 1939

7.E-mail from Dale Drinnon to Richard Muirhead December 20th 2010

8.The Argus July 25th 1911

9. The Argus August 8th 1911

10. Dr C. Fenner Bunyips and Billabongs (Sydney: Angus and Robertson,1933) in Dr K.P.N. Shuker Extraordinary Animals Revisited (Bideford: CFZ Press, 2007) p. 160

Apparitions in the Air, Cheshire, 1651

By A.J.F.

The following article appeared in The Cheshire Sheaf, vol 56 May, 1961, pages 40-41

Apparitions in the air in 1651

The following strange pamphlet is taken from the Thomason Tracts in the British Museum E.628.16. It is written in the form of a letter dated 29 April 1651, and purporting to be from one William Radmore, to his Brother.

"Wonderful News from the **North:** Being a true and perfect Relation of severall strange and Wonderful Apparitions seen in the Ayr, between Madely and Whitmore, in the County Palatine of Chester

This Relation was taken by the Minister of Madely,and the truth thereof verified by Mrs. Holt of Oakershill, who with her Maid, were eye witnesses thereof.

London. Printed for George Horton. 1651

" I shall relate to you, a strange Apparition that was seen near to us, on Wednesday the II of this instant, at a place called the Highwayhouse, between Madely and Whitmore: The Woman to whom it appeared, was and is a Religious Woman, and come of a godly Family, the manner thus: On the said day the Woman Sitting in the door with a little Girl in her arms, she perceived the Sun to shine exceeding red, and casting her eyes upwards, she beheld a dark body over the sun, about the bigness of a half moon, and in a short space , the said body divided into several parts, seeming numberless to her view, about the bigness of small Pewter dishes, which came swiftly towards her and immediately the Court about the house seemed to be filled with armed Hands and Gauntlets, with swords; glittering and fighting (in their imagination) with another as great an Army, and it seemed to their view to be in the ayr above them.

At the site whereof, she was amazed, and being greatly astonished, ran into the house, having left behinde her another little Girle playing forth of the doors, her Maid and she presently betaking themselves to prayer, and after receiving some encouragement, they opened the door, and the Maid boldly stept forth and took in the child (to her supposition from amongst them) which had no harm; But the numbers so encreased on both parties, that the House became all darkened like night: Then when she got in her childe, she went to prayer again, verily supposing her end to be near at hand; After which her Maid and she looking forth, beheld infinite of Horse legs and feet trampling, and great Canons and Ordnance on the other side of the House rear'd up together, with their muzzles upwards, and the Houlsters hanging on them; this continued some space and vanished. Then she opened the door and went forth, and saw the likeness of one man onely, standing within the Court near to the mote, and immediately there arose a little Cloud or Vapour (as it were) forth of the mote, from whence issued a Bird about the bigness of a Capon, with wings such as an Angel is usually portraicted with, and a second and third followed and flew near to her and her Maid, having faces almost like Owls, and something resembling a man, and after they had hovered about a while, they vanished in vapour or Cloud; These birds were of a blewish-red, or sanguine colour, but the Men, Horse, Swords, and Canons, all like fire.

I had almost forgot one passage, her Maid at first saw nothing, but after she had uttered these words, Mistris it may be the Lord wil not suffer me to behold what you do, she immediately saw the same. Oh who would not fear and love thee thou King of Saints, how unsearchable are thy ways, to poor hoodwinckt mortals......

A.J.F

Ezekiel's Wheels

by GOD and Richard Muirhead

Fig 1 Ezekiel's Wheels

Reproduced with the kind permission
of Rolf Stark of Biblical-art.com from " Die Bilder zur Bibel"
(Peter Meinhold, 1965)

Ezekiel chapter 1 verses 15-21 " As I looked at the living creatures, I saw a wheel on the ground beside each creature with its four faces. This was the appearance and structure of the wheels: They sparkled like chrysolite, and all four looked alike. Each appeared to be made like a wheel intersecting a wheel. As they moved, they would go in any one of the four directions the creatures faced; the wheels did not turn about as

18

the creatures went. Their rims were high and awesome, and all four rims were full of eyes all around."

"When the living creatures moved, the wheels beside them moved; and when the living creatures rose from the ground, the wheels also rose. Wherever the spirit would go, they would go, and the wheels would rise along with them, because the spirit of the living creatures was in the wheels. When the creatures moved, they also moved; when the creatures stood still, they also stood still; and when the creatures rose from the ground, the wheels rose along with them, because the spirit of the living creatures was in the wheels. "

This is about God not an U.F.O.

For a Christian approach to U.F.O.s see http://www.be-ready.org/ufos.html

Flying Snakes Part 1

by Dale Drinnon

This blog originally appeared in Cryptozoology Online

on January 23rd 2011

Oll Lewis from the information on cryptozoology.He blog post a while Welsh cryptids specifically a snake or dragonet of Glamorgan.Gwibers Wales in folklore CFZ is a font of Welsh did an excellent back on one of the named the **Gwiber**, variety of flying South are reported all over but it was in Penllyn they are reported more recently in Marie Trevelyn`s *Folklore and Folkstories of Wales* from 1909:

"The woods around Penllyne Castle,Glamorgan, had the reputation of being frequented by winged serpents,and these were the terror of old and young alike. An aged inhabitant of Penllyne who died a few years ago, said that in his boyhood the winged serpents were described as very beautiful. They were coiled when in repose, and " looked as though they were covered in jewels of all sorts.Some of them had crests sparkling with all the colours of the rainbow. When disturbed they glided swiftly, "sparkling all over," to their hiding places. When angry, they " flew over people`s heads, with outspread wings bright and sometimes with eyes,too,like the feathers in a peacock`s tail. " He said it was " no old story," invented to "frighten children", but a real fact. His father and uncles had killed some of them, for they were "as bad as foxes for poultry" This old man attributed the extinction of winged serpents to the fact that they were " terrors in the farmyards and coverts.""

Oll goes into some depth on the mystery in his post and he did attempt some interviews. His conclusion: the creatures are reported as unlikely to exist. Some facts do stand out from the traditions: one is that a feathered skin from one of these flying serpents was kept by one family for many years. Whatever else might be said, a feathered skin is a real object and feathers necessarily mean the creature was a **bird**. Another feature is that these creatures could sometimes be seen seemingly "coiled up". Several long-necked birds sleep with the head and neck curled back towards the tail and some long tailed birds also throw the tail around the front. Furthermore, these flying serpents are often said to have clawed feet and the claws are also said to be poisonous. `**Gwiber**` means `viper` and is the same as the French `voivre` from which we derive the term `wvre` or **wyvern.**

A wyvern is a smallish two-legged and winged dragon, and wyverns are traditional over many parts of Western Europe. They may have their exact counterparts in Eastern Europe in the aitavars and other creatures that are simultaneously `dragons` and like barnyard fowl, sometimes described as having tails of fire, some `firedrakes` and perhaps the Russian firebirds. If this is so, then we seem to have two distinctive populations where the gaudy males are divided by their coloration, a Western branch which is primarily green, and an Eastern branch where the males are primarily red. And the Welsh folklore, when speaking of the flying serpents describes their feathers as peacock-like. That does also determine what kind of creature they really are- they are pheasants (peacocks and domestic fowl are also related to pheasants.)

Since I had already seen where the anhinga was described as a flying snake because of its long neck, I assumed that the gwiber or wyvern was a sort of a large pheasant with a very large neck and a very large tail. It might also be the same as a **cockatrice, which Wikipedia** describes as a sort of a fowl with a long lizard like (snake like or dragon-like) tail. It is said to be particularly vicious and is said to have a venomous breath (or a venomous bite, or venomous claws, or a lethal gaze); any of those descriptions could be probably taken as awful warnings that people should keep well away from them, but they need not be true: people are always saying any number of perfectly harmless animals are venomous, especially when they are snakes. Believing that the long-necked and long-tailed pheasant was a viper would just about be typical: the flying serpent reports that turned out to be anhingas also insisted on their potent venom

The Welsh flying snakes are said to be quite aggressive and to kill poultry when given a chance. They will also attack travellers and they are said to be roused to fury at the sight of a red cloth. Reports in this general category of flying snakes commonly put the length down as from six or nine feet.

The largest kind of pheasant is the Reeve's pheasant, native to China. It has a very long tail and can regularly grow to over six feet long, up to eight feet long. The size is in the right range, and adding a long snake-like neck to such a bird would make it even longer. It is a hardy bird able to stand extremes of cold and heat, and the males are said to be hostile to humans, dogs, and especially to males of other pheasant types. If they are being raised together, Reeve's cocks must be kept . separate from the males of other kinds because of this aggressive nature

The hen Reeve's pheasant is as large as the male of the common (ring-necked) pheasant and already has a fairly long tail of its own. This could lead to the tale that cockatrices arise from eggs laid from roosters; peasants unfamiliar with the bird might well mistake the hen for the rooster of another species.

END OF PART ONE

FIGURES

1. Wyvern and Cockatrice from Church, Exeter.

2. Flying Serpent from Deviant Art.

3. Wyvern

4. Pheasants to scale from Wikipedia

5. Cock and Hen Reeve's Pheasants. Copyright Dale Drinnon.

The Giant Centipedes of Hong Kong

by Richard Muirhead and Mike Hardcastle

According to scientific orthodoxy, in the form of `Hong Kong Animals` by Dennis S.Hill and Karen Phillipps there are 4 species of centipede in Hong Kong. These are in ascending order of size: The Long-legged Centipede Thereupoda clunifera, the Smaller Forest Centipede Scolopendra morsitans , the Urban Giant Centipede Scolopendra multidens and finally S.dehanni. This short essay will be concerned with S.multidens in particular, extraordinary large specimens of which have turned up in Hong Kong over the last 100 years. There was also a species of centipede in Hong Kong called Scutigera at the time the naturalist Geoffrey Herklots`s `*The Hong Kong Countryside throughout the seasons*` was published in 1951 but according to Max Blake of the Centre for Fortean Zoology:

"To my knowledge no Scutigera is found in Hong Kong" Max also said " Well S.gigas doesn`t actually exist itself, it is a synonym of S.gigantea. As far as I can see, the two species are totally separate coming from Venezuela and S.multidens coming from around Taiwan and neighbouring countries. "(1)

Geoffrey Herklots wrote in the ` *The Hong Kong Countryside*`: concerning the Scutigera:

"There is another large centipede found in Hong Kong, and in other parts of the tropics called *Scutigera,* the shield-bearer, which scuttles or runs very quickly. These have very long legs which lift the body well above the ground and they run about during the day as well as at night in search of small insects on which they feed. A friend of mine calls these scuttlers, emperor centipedes. They are much feared by the Chinese who say that their bite is fatal to man; this is of course quite untrue."(2)

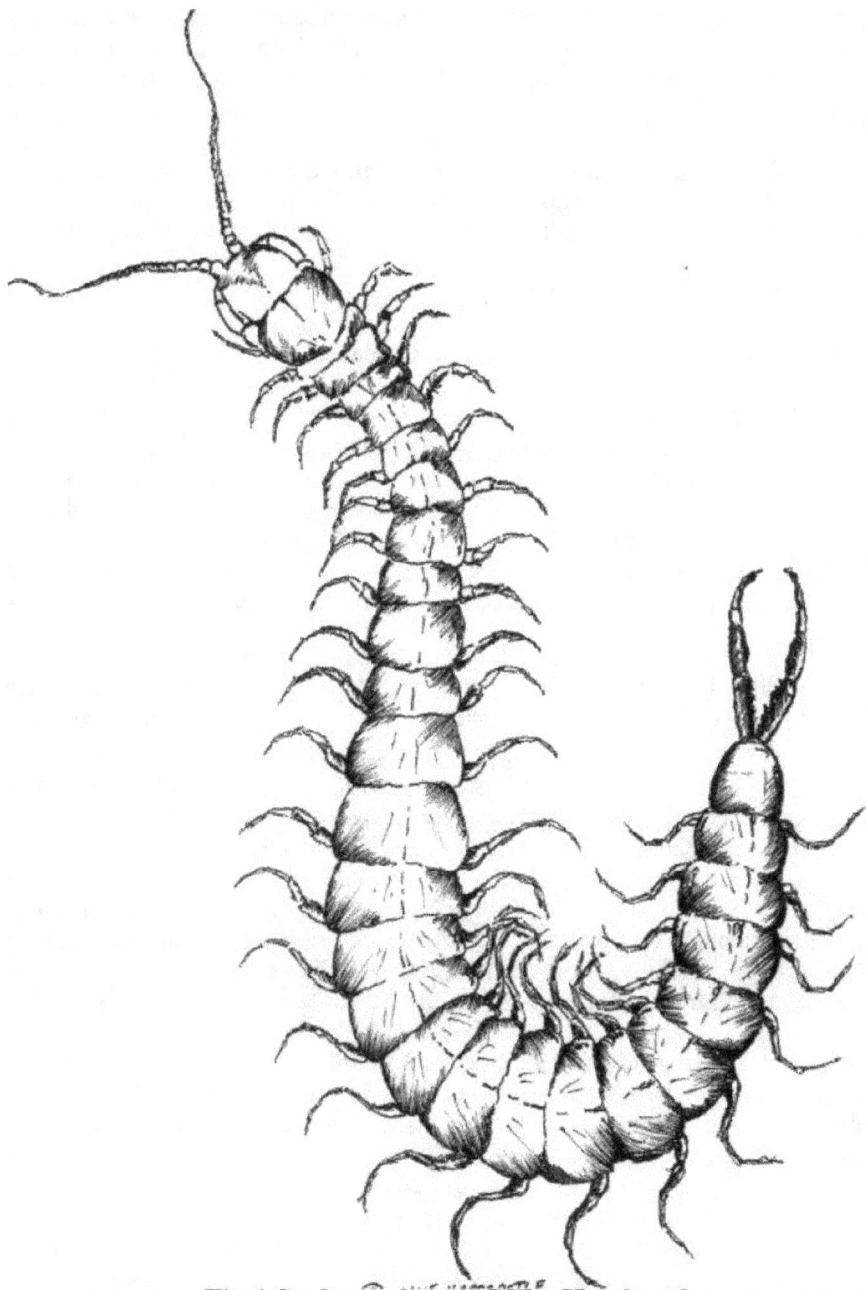

Fig 1 Scolopendra sp. Mike Hardcastle.

25

Another mystery is the presence of Scolopendra subspinipes in Hong Kong, but as far as I know this species is not now in the former Colony. This species is also known as the Vietnamese centipede. (3) However S.multidens is a sub-species of S.subspinipes.

Figure 1 shows Mike Hardcastle`s representation of Scolopendra based upon an image on the Internet. Based upon web sites including a Flickr phot at http://www.flickr.com/photos/lat3ralus/4972286219/in/pool-1456628@N20/

Hill and Phillipps say about S.multidens:

" *Scolopendra multidens* (Urban Giant Centipede)

A large brown centipede, this `urban giant` measures 10-13cm in body length, and has a body composed of a series of long and short segments. Each leg is tipped with a single sharp spine. Under the head is a pair of large black-tipped poison fangs. The bite of a small species is painfull so it is assumed that the bite of a large *Scolopendra* is very painful.They are reputed to be able to kill small mice with their poison, although it is to be expected that their normal prey would be American Cockroach. If handled, they are very quick to bite."

The other large local species is S.dehanni, which is identifiable by having only three spines on the last pair of legs as opposed to the seven spines on each last leg of S.multidens (five dorsal and two ventral) Also, it is ecologically different in being a forest litter species. (4)

In my Muirhead`s Mysteries blog in `*Cryptozoology Online*` of September 21st 2010 `Giant centipedes in Hong Kong` I wrote the following, reproduced with the permission of Jon Downes, of the CFZ, very slightly amended.

"Anyone who has lived in Hong Kong, especially any entomologist or Fortean, will be aware of the centipedes, Scolopendra spp, sometimes red in colour that live in many parts of the former British territory. They certainly lived on the Peak, where Jon and I lived as youngsters from the 1960s-1980s.

It is known that these myriapods can attain a length of nearly 30cm/1ft. However, there are claims that these centipedes can grow to 60cm long; that is 2ft! (Thank you to Heywood`s Mathematical Genius Lizzy for this conversion; it took me several years to get O Level Maths).

In one of those bizarre Fortean incidents, at the end of last week I was browsing the online Hong Kong newspaper archive, which like some quasi-bibliographical giant, rarely sleeps in this house, and I accidentally typed in the words GIANT CENTIPEDE in capitals thus, when 99% of the time I type the words in lower case, and I came up with the most interesting of the total of only 7 hits, dated June 30th 1924, front page of the *Hong Kong Telegraph* http://hkclweb.hkpl.gov.hk/hkclr2/internet/eng/html/frm-bas_srch.html (This is the website for the search engine not this issue of the HK Telegraph.)

There is a very similar story from March 19th 1948 this time from *The China Mail*.....it read:

" 2-Foot Long Centipede A giant centipede, almost two feet long, was caught and destroyed in the 3rd floor kitchen of No 80 Fuk Wa Street Kowloon, [N.W.Kowloon-Richard] on Tuesday [March 16th-R]. The large insect was discovered near the fireplace and killed with a piece of firewood by one of the inmates of the flat" [Inmates,that`s an interesting choice of words, what was he a prisoner?-R] Interestingly the 1948 centipede was also found in Kowloon, that part of the mainland directly across the harbour from Hong Kong."(5)

The 1924 item mentions a company called Messrs A.S. Watson & Co in Kowloon. This company still exists and has offices all over Hong Kong including Kowloon Tong which is several miles east of Fuk Wa St.

There is on the Web a video of a giant centipede Scolopendra multidens/S.dehanni on Cheung Chau Giant centipedes (Scolopendra multidens/S.dehanni) are surely Hong Kong`s ultimate creepy-crawlies: not only do they look spooky and grow to 13cm long, they run fast and pack venom they inject through fangs (6) Cheung Chau is an island to the west-south-west of Hong Kong island. I also recall a Facebook contact describe a centipede on The Peak as being c.20cm, in length, but 60cm? How big do centipedes grow?

The 2ft specimen from 1924 is described as S.gigas.

There is an interesting story in *The Hong Kong Daily Press* July 1st 1924, perhaps the editor had the story in his rival paper of the day before in mind (see above). As follows:

SWALLOWED A LIVE CENTIPEDE
AN ANTI-TOXIN: CHINESE WOMAN'S ORDEAL.

A correspondent of the *North China Daily News* writing from Mienchow, Szechuan on June 9th says: " A few days ago a Chinese woman in the city was extremely ill. She went to a local doctor and he prescribed a centipede, charging her three dollars, and instructing her to eat it. It was a poisonous variety and the idea seems to have been that act as an anti-toxin to the poison in her system. It made her very ill. Her face and lips swelled to such an extent that she ate no food for five days.

This is only one of many examples that might be given which show that the ordinary Chinese is still living in a world that belongs to a far distant past." (7)

[that would have been a classic quote for my *China: A Yellow Peril?* book!]

The Hong Kong Telegraph article of June 30th 1924 is as follows:

A GIANT CENTIPEDE

INTERESTING CAPTURE
AT KOWLOON

How big does a centipede grow? Those persons whose knowledge of this creature of many legs does not extend beyond the species found in temperate zones will probably reply " Only a few inches". But those who have lived for some time in tropic lands will have seen, or heard of centipedes of eight inches or more in length. *Scolopendra gigas,* as its name implies, is the giant of all the centipede genera and species, and attains a length of twelve inches. It is found in the warm regions of Asia, and local residents will be interested to know that one of these gigantic representatives of the tribe has been captured in the heart of Kowloon's residential area-whether a native of the place, or brought originally in timber from further south, cannot yet be said.
Last Saturday night, Mr J.Gibson of Messrs. A.S.Watson & Co's dispensary in Kowloon heard what he took to be mice scampering along the floor of a room over the dispensary, where workmen have recently been carrying out certain repairs.

Mr Gibson investigated, and found a huge centipede crossing the room. Anxious not to damage it, he looked around for some means of capturing it alive. His eye lighted on a soup plate, and at some risk to himself he placed the plate over the creature and had it imprisoned. Going down to the dispensary,he procured some chloroform, and poured a small pool of the drug on the drug on the floor beside the plate. Maneouvring this, he draw it over the chloroform, with the centipede still underneath, and the anesthetic put the quietus on the struggling *Scolopendra*.

The centipede now reposes in a bottle of alchohol, in the dispensary, and although it has shrunk slightly through immersion in the spirit it is still over ten inches in length. When alive, Mr Gibson estimates, it must have been quite a foot long. This monster centipede has two wicked looking calliper jaws, and twenty-three pairs of legs.

The term "centipede" is misleading, for these creatures have a varying number of legs, ranging from fifteen pairs in some species to one hundred and seventy-three pairs in others. The allied " millipedes" are all not necessarily blessed with a thousand feet.

The bite of most tropical species of centipedes, such as Scolopendra, is painful, and under certain circumstances even dangerous. Some kinds of centipedes are harmless, like the slender,phosphorescent ones seen on walls and sometimes erroneously termed "earwigs" under the impression that they creep into a sleeper's ears. Centipedes are related to insects but (with millipedes) belong to a class by themselves the Myriapoda, which is a self-explanatory term. They lurk beneath stones, or in dark places in houses and come forth at night in search of small insects. (8)

REFERENCES

1. E-mail from Max Blake to Richard Muirhead February 6th 2011

2. G.Herklots The Hong Kong Countryside (Hong Kong:various sources1860-1952) p.?

3. G.Herklots Ibid p.138 and E-mail from Mike Hardcastle February 25th 2011

4. D.S.Hill and K.Phillipps Hong Kong Animals (Hong Kong: Government Printer, 1981) p.218

5. The China Mail March 19th 1948

6. Giant centipede on Cheung Chau http://www.hkoutdoors.com/videos/giant-centipede-on-cheung-chau Accessed 20/09/2010

7.The Hong Kong Daily Press July 1st 1924

8.The Hong Kong Telegraph June 30th 1924

The South Shields Devil Crabs

by Mike Hallowell

Two years ago, my book *Mystery Animals of the British Isles: Northumberland & Tyneside* (CFZ Press, 2008) informed readers about a cryptid called the Cleadon Big Cat, a terrifying sea monster at Marsden Bay called the Shony and a fascinating entity known as the Giant Lobster of Trow Rocks. Since then I've written up stories about a desiccated mermaid on display in a barber's shop, an entity called the Horse-Man of Bede and a weird man-beast from Hebburn known as Blue Eyes, not to mention a huge, hairy hominid said to wander Cleadon Hills after dark.

Now South Tyneside is the smallest Metropolitan Borough in the country, and you'd think its residents would be pleased to brag about one cryptid, let alone seven. However, against all the odds it seems another cryptozoological conundrum may have to be added to our rich and varied folklore.

Two years ago whilst chatting to Ronan Coghlan, I purchased a copy of his *book A Dictionary of Cryptozoology*, (Xiphos Books, 2004) and have to confess its one of the most fascinating tomes I've ever come across. Now here's the funny bit; I fascinated myself by imbibing strange tales of the Antarctic Narwhal, the Hairy Fish and the Sherwood Forest Thing. Then, without warning, my eyes were drawn to a short entry entitled, *South Shields Crab*. As I only live a very short distance from South Shields, and have written literally hundreds of articles concerning its Fortean history, I must be forgiven for becoming somewhat excited.

According to Ronan, the possibility exists that a hitherto unrecognised species of crab might be living off our coastline, although he does acknowledge that it could just possibly be, "a colour variation of a known species".

Well, I've heard tales about these mystery crabs before, and they fascinate me.

Some years ago I had several engaging conversations with the late archaeologist Evelyn Waugh-Almond (she was alive then, for the record) and she told me that just off the northerly aspect of Marsden Bay, at the rocky outcrop known as Velvet Beds or Camel Island, there were "crabs living unknown to man".

Now back in Victorian times, Velvet Beds was a favourite pic-nic spot. Hordes of mothers, fathers and their offspring would go there with meat pies, ham sandwiches and tubs of potted brawn to take in the sea air, which was said to be most efficacious in the treatment of the humours and, if you were unfortunate enough to have them, the vapours.

At that time the rock was covered in a thick carpet of lush, dark green grass which supposedly felt "just like a bed of velvet" under one's feet. According to tradition, that's how the rock came to be known as Velvet Beds. The grass has all but gone now – only a few tufts remain – and most folk refer to the rock as Camel Island due to the fact that rapid erosion of the striated Magnesian limestone has left it looking like a camel's hump.

Velvet Beds, or Camel Island allegedly the home of the South Shields Devil Crabs

Image © Thunderbird Craft & Media 2008

But there's another tradition, which espouses the idea that the rock gained its name from the large number of velvet crabs which inhabited the waters around it.

Evelyn told me that the crabs "unknown to man" looked like velvet crabs, but were taxonomically different. They were alleged to have a "nasty disposition" and were extremely aggressive. This, plus their distinctive red eyes – also possessed by velvet crabs, I've been told – led to them being given the alternative monikers of Devils Crabs and Witches Crabs.

One correspondent told me that the crabs at Velvet Beds can grow to a width of 14 inches, which makes them far larger than the average velvet crab. To my knowledge, none of this size have ever been caught. Trevor Wilkinson, another reader of my *WraithScape* newspaper column, told me that they can grow to "enormous size". Just how enormous he was unable to say.

Rowan references *Animals and Men* as his source for the story, but doesn't give a particular month or year or provide the issue number, so I'm hoping Jon Downes might be able to provide some more detail on this cryptozoological enigma. I've put out a call to all South Tyneside's *craberati*, hoping that someone might come forth with a photograph, a specimen or at least an anecdotal tale or two.

The Pink-Tusked Elephants of Tang Dynasty China

by Richard Muirhead

In early 1999 I began investigating the black elephants which inhabited the districts of Hsün and Lei, China in T`ang Dynasty China 618-907 A.D. There is even one report of them being alive at the "beginning of the 19th century" according to `Celestial lancets: a history and rationale of acupuncture and moxa` (1) by Gwei-Djen Lu, Joseph Needham and V.Lo. Dr Karl Shuker noticed my query which was on cz@onelist.com and summarized my research in his `Alien Zoo` column in `Fortean Times` 122 May 1999 :

Think Pink

"The onelist cryptotozoology discussion group has yielded some fascinating snippets of previously overlooked or little publicised data since its establishment last year. One the latest stems from the online revelations of cryptozoological researcher Richard Muirhead that Edward Schafer`s book `The Vermilion Bird` (1967), concerning life in T`ang Dynasty China (618 – 907 AD), refers to a race of black elephants in Hsün and Lei, corresponding to the Leizhou Peninsula and southeastern Guangxi Province. It appears that this peculiar form of pachyderm has been formally dubbed Elephas maximus rubridens by zoologist Dr P.E.P Deranagala, who used as his type specimen a depiction published in 1925 of an antique Chinese bronze statuette, held at Chicago Field Museum of Natural History. Can Fortean Times readers supply any further details re these elephantine enigmas? If so, you know where to write. Assorted cz@onelist communications,Fortean Times Feb-Mar 1999 (2) "

As far as I know Karl didn`t receive any further significant information about Elephas maximus rubridens, nor did I.. In `The Vermilion Bird` Schafer said: "Herds of wild elephants trampled the cultivated fields of Honan and Hupeh in the fifth Christian century, and were not a rarity in Huai- nan in the sixth.(3) "In T`ang times, they still roamed the woodlands of Nam-Viet, and were even abundant in the northern counties of Ch`ao and Hsün." (4) "One T`ang source tells of a race of black elephants with small pink tusks in Hsün and Lei." (5) "Perhaps this describes the true Chinese race itself, whose furious representatives had been subdued by the agents of the kings of Shang.In any case, the peoples of Nam-Viet caught and killed them with poisoned arrows in T`ang times, and roasted their trunks to make delicacies for tropical feasts, (6) the tusks, or some of them, were sent to northern artisans for conversion into chopsticks, hairpins, combs, plectrums, footrules, note tablets, and for inlays in fancy cabinet work, the ivory being dyed in a variety of colours." (7). "The naturally pink ivory of the local elephants was well favoured, indeed regarded as equal to the ivory imported from overseas, but the bulk of the regular tribute ivory for the use of the court came from Hunan-chou in Annam. (8) I know nothing of its natural colour".

Image © Thunderbird Craft & Media 2010

In 2010 I obtained a copy of Berthold Laufer`s ` Ivory in China` from the Bodleian, Oxford: Laufer said: "Ivory occupies a very prominent place in the art of the Far East, and Chinese carvers in ivory have always stood in the front rank of their craft....(9) The archaeology of ivory and the older real works of art created in the substance have almost wholly been neglected." Later Laufer said, of the twelfth century: " The tusks imported by the Arabs [to China-Richard] are described by a contemporary observer as being straight and of a clear, white colour, with patterns displaying delicate lines. In weight they varied from fifty to a hundred pounds, whereas the tusks coming from Tonking and Camboja were small, weighing only from ten to twenty or thirty pounds, and had a reddish tint." (10)

John Moore in a posting on cz@onelist.com on February 28th 1999 said:

" Those interested in this animal may wish to check out the work of P.E.P Deraniyagala, who named what appears to be the same animal Elephas rubridens on the basis of textual and artistic evidence (Deraniyagala, Proceedings of the Fifth Annual Session, Ceylon Association of Science pt 3 p. 10 1950) ; see also the same author, Spolia Zeylanica 26 (1951) : pp 50-51 and Some Extinct Elephants, Their Relatives and the Two Living Species (1955), pp 124-125. Deryagala`s source for the information on this elephant is B.Laufer, 1925, Ivory in China. Field Museum of Natural History Anthroplogy Leaflet 21." (11)

Shoshani and Tassy (The Proboscidea p.370) list Elephas maximus rubridens as a synonym of E.maximus but do not discuss it further (12)

Someone calling himself/herself `Blackhawk` sent an e-mail on the same date to cz@onelist.com saying:

"Just curious, but was Deraniyagala's identification of the beast accepted under the rules of the International Code of Zoological Nomenclature? If it was how did he get past the rule of thumb requiring at least a holotype in some form or the other." (13)

The Internet is not always accurate as we all know, including on matters elephantine. Witness the following statement on the Civilization Fanatics Forum, a gaming forum, about Elephas maximus rubridens:

" Today, only two main species exist, but in the time of the Nabateans [3rd century B.C to at least 106 AD - Richard] four separate species were known at that time, and could have been used for warfare. One thousand years earlier, several races of Asian and African elephants had become extinct (about 1500 B.C.) For example the Elephas maximus rubridens existed in China as far north as Anyang, in northern Honan province. Writings from the 14th century B.C. state that elephants were still to be found in Kwangsi Province. " (14)

Dale Drinnon has stated:

"I believe you sent me an inquiry an inquiry about the pink-tusked elephant........My opinion is that actually the reported colouration cannot be trusted, it might originally have been only an artistic convention, and that otherwise, unusual colouration is a very dubious indicator for new species. Many Cryptozoolologists attach too great an importance to that. There are a good many putative Cryptid variations on the Asiatic elephant but so far as we can tell, all of them are only rare variations of one overall species. " (15)

REFERENCES

1. G. Djen-Lu, J.Needham, V.Lo Celestial lancets: a history and rationale of acupuncture and moxa (London: Routledge Curzon, 2002) p. 3172. K.Shuker Alien Zoo Fortean Times 199 May 1999. p. 19

3. E.Schafer The Vermillion Bird T`ang Images of the South (Berkeley and Los Angeles: University of California Press, 1967) p. 224 citing E.Schafer War Elephants in Ancient and Medieval China,Oriens , Vol 10 (1957) pp 289-291

4. E. Schafer Ibid p 224 citing Liu Hsün Ling piao lu b 6

5. E.Schafer Ibid pp 224-225 citing Tuan Kung-lu Pei hu lu 2 8a

6. E.Schafer Ibid p.225 citing Liu Hsün Ling piao lu b 6; Pei hu lu ,2 ,8a-8b; Tai Chün fu , Kuang I chi in T`ai p`ing kuang chi, 441, 4a

7. E.Schafer The Golden Peaches of Samarkand: A Study of T`ang Exotics (Berkeley and Los Angeles: University of California Press, 1963) p.240

8. E.Schafer op cit p. 225 citing Pei hu lu ,2, 8a-8b; Li Chi fu, Yüan ho chün hsien (t`u) chih 38, 1086; E.Schafer and B.Wallacker Local Tribute Products of the T`ang Dynasty Journal of Oriental Studies Vol 4 (1957-1958) pp 213-248

9. B.Laufer Ivory in China. Anthropolgy Leaflet No. 1 (Chicago:Field Museum of Natural History,1925) p.1

10. B.Laufer Ibid p.17

11. E-mail from J.Moore to cz@onelist.com February 28th 1999

12. Shoshani and Tassy The Proboscidea Evolution and Palaeoecology of Elephants and Their Relatives (Oxford: Oxford University Press, 1996) p.370

13. E-mail from `Blackhawk` to cz@onelist.com February 28th 1999.

14. Rambuchan`s posting on Civilization Fanatics Forum website September 14th 2005

15. E-mail from Dale Drinnon to Richard Muirhead January 11th 2011

The Israeli Mermaid

by Zvi Ron

The world was astounded to hear reports from Israel about a mermaid sighted on the beach of Kiryat Yam, a town north of Haifa. The sightings began in August 2009 and were widely reported all over the world. Most accounts followed the original Jerusalem Post article . (1) The article stated that "dozens of sightings" had been reported over the past few months. Town council spokesman Natti Zilberman is a central figure in all newspaper articles on this subject. He is quoted as saying, "Many people are telling us they are sure they've seen a mermaid and they are all independent of each other," and "People say it is half girl - half fish, jumping like a dolphin. It does all kinds of tricks, then disappears." Shlomo Cohen, a retired career soldier, reported that he and five friends saw a woman who, "was laying on the sand in a weird way" and jumped into the water upon being spotted, it was then that they noticed that "she had a tail." (2) A thirteen year old boy named Uri told reporters of his mermaid sighting, describing a creature with a fin and long satin hair, but spiky in front. She was too far away so Uri did not see her face and could not tell the reporter if she was beautiful. (3) Zilberman offered a million dollar reward to anyone who can provide proof of the existence of the Kiryat Yam mermaid. When asked if the council of Kiryat Yam has the money to pay the reward Zilberman explained "I believe, if there really is a mermaid, then so many people and tourists will come to Kiryat Yam, a lot more money will be made than one million dollars." (4)

While insisting that it was not a publicity stunt, the mermaid reports were widely perceived as such. Kiryat Yam residents had long dreamed of something that would "transform it from an unemployment-cursed and crime-ridden blue-collar town of 45,000 residents into a flourishing resort". (5) The mermaid reports did in fact put Kiryat Yam on the tourist map, attracting many people to the town's beach, hoping for a glimpse of the mermaid and a chance at the reward money. (6) As a result of the increased tourism, Zilberman announced that the town would soon open its first hotel. (7) A new boardwalk was constructed along the beach, displaying a statue of a mermaid, sent from Hungary. Reports surfaced that the municipality was attempting to sell "Mermaid Beach" to investors for three million dollars, something which is against the law. (8) When reporters asked Zilberman for the contact information of the eyewitnesses, he replied that they do not have cell phones but if the reporters would visit Kiryat Yam in person, the witnesses will be happy to discuss what they saw. (9)

This is not the first time council spokesman Natti Zilberman has been involved in what seems to be a publicity stunt aimed at attracting tourists to his town. He had previously announced that the town would import lion dung from the Ramat Gan safari to scare off wild horses in the vicinity and that local teens would be collecting jellyfish from the beach in order to sell to workers from China who consider jellyfish a delicacy. At a press conference, Zilberman claimed that jellyfish sales would add fifty thousand shekels to the town's coffers annually. This announcement was greeted with laughter from the assembled journalists. The importing of lion dung and mass jellyfish captures never took place, but served to keep Kiryat Yam in the news for some time. (10)

But there may be somewhat more substance to the mermaid sightings than Zilberman's reputation would lead us to believe. The mermaid reports attracted a documentary crew from the United States television network NBC. The film crew spent a whole week on the beach of Kiryat Yam and filmed morning and night, both underwater and above it. The crew claimed that during one of the late night outings they managed to spot a human figure dipping in the water – and then disappearing underwater. The show's researchers hurried and dived after the figure, but were unable to trace it. The show's findings, along with the footage shot by several bystanders, were transferred to the Center for Coastal Ocean Research in Los Angeles. The center's director, Michael Shacht, examined the evidence and said that it was impossible to unequivocally determine that the figure in the footage was indeed a mermaid. According to Shacht, the team might have succeeded in capturing the rare phenomenon on tape. The investigative report received many responses in the United States and drew a lot of interest among viewers and ocean and sea researchers around the world. (11) There has been no follow up reported since then regarding this documentary.

It did not take long for the Kiryat Yam mermaid to be co-opted for various political and ideological purposes. People for the Ethical Treatment of Animals (PETA) announced that they would feature the Kiryat Yam mermaid in a billboard campaign to encourage vegetarianism. (12) An American organization claiming to defend the rights of mermaids threatened to appeal to the International Court of Justice in The Hague against the town of Kiryat Yam over the prize money offered for evidence of the mermaid. The organization, presenting itself as the Mermaid Medical Association in Brooklyn, New York, claimed to be shocked to hear about the prize offered by the town, saying it "badly and outrageously damages the legendary mermaid legacy." The organization informed the municipality that it had 10 days to take back its announcement of the prize, or they would approach the International Court of Justice in the Netherlands and demand that it intervene. (13) Though reported uncritically by many media outlets, the lawsuit turned out to be a hoax. The Mermaid Medical Association turned out to be a health clinic on Mermaid Ave. in Brooklyn, New York. (14) The receptionist there was confused when asked about the letter sent to Kiryat Yam, claiming to know nothing at all about the matter. "Do we even have mermaids?" she asked. "I've never seen one." The entire story was then used to demonstrate the lack of proper investigation when dealing with news from the Middle East. (15) It is probable that the lawsuit hoax was initiated to poke fun at the constant barrage of attacks against Israel on human rights grounds, perceived by many Israelis to be knee-jerk responses to just about anything that takes place in Israel.

Is the Kiryat mermaid real? For the residents of Kiryat Yam the mermaid has generated an influx of tourism, investment and interest. What could be more real than that? has generated an influx of tourism, investment and interest. What could be more real than that?

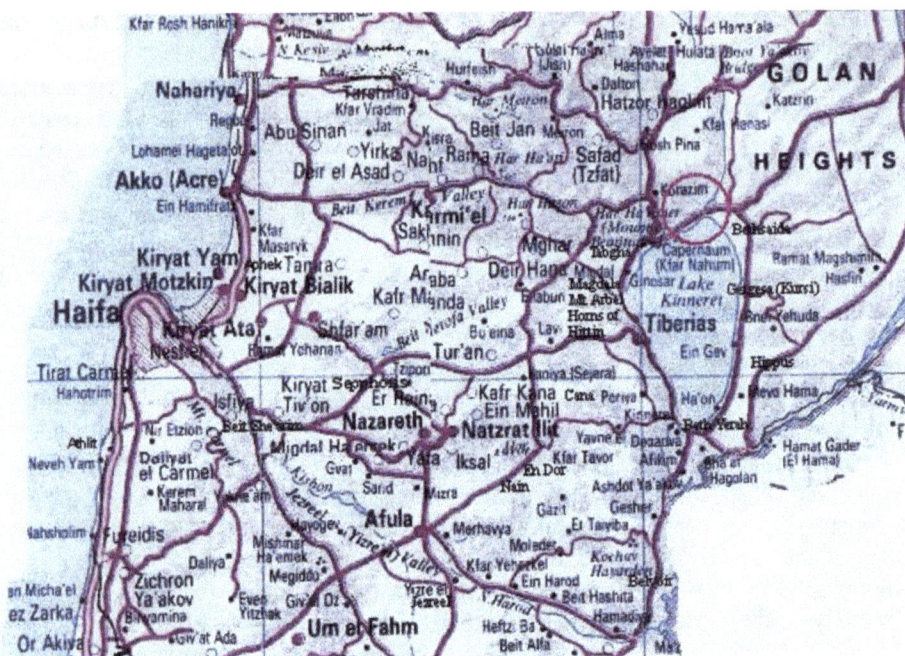

Fig 1   Map of Northern Israel.

Kiryat Yam is just up the coast from Haifa. See
http://www.alamagorhouse.com/images/mapisrael.jpg

REFERENCES

1. Jerusalem Post, "Mermaid sightings reported in Kiryat Yam", August 11,2009.

2. Arutz Sheva News, "Mermaid spotted on Kiryat Yam beach", August 12, 2009.

3. Haaretz, "How the mermaid changed the lives of the residents of Kiryat Yam", September 25,2009.

4. Jerusalem Post, "Mermaid sightings reported in Kiryat Yam", August 11, 2009.

5. Haaretz, "Tough blue-collar town wants to become a prime tourist destination," February 21, 2008.

6. Haaretz, "Is a mermaid living under the sea in Northern Israel?", August 12, 2009

7. Nrg.org, "Kiryat Yam: Still looking for the mermaid," November 17, 2009

8. Channel 2 News, "Kiryat Yam is selling " Mermaid Beach" unlawfully", December 6, 2009

9. Haaretz, "How the mermaid changed the lives of the residents of Kiryat Yam", September 25, 2009.

10. Chadashot HaKriyot, "Fairy tales from Kiryat Yam", issue 209, August 2010

11. Ynet, "NBC: Kiryat Yam mermaid might be real," April 5, 2010.

12. Jerusalem Post, "PETA plans Kiryat Yam billboard featured a topless mermaid," August 26, 2009.

13. Ynet, "Kiryat Yam to be sued over mermaid?", August 24, 2009.

14. New Jersey Jewish News, "A (half-woman half-) fish tale", August 25, 2009.

15. New Jersey Jewish News, "Reporting sinks mermaid story," August 26, 2009

A Wildcat from Shaftesbury, Dorset

by Andy Scott

Whilst on holiday in Dorset (the last week of May 2010) my partner and I walked in a quiet park (called a breathing space) in Shaftesbury which looked out onto an expanse of marshland/meadow. Looking through binoculars I noted a pheasant and a rabbit, then as I scrolled right towards the edge of the meadow near to the woodland I spotted a big cat with dark tabby-like markings, it immediately struck me how big it was from a distance of about ¼ of a mile,its tail was thicker than a domestic cat and as I observed it, the cat turned its head and I could see the wild stare in its eyes as it looked straight into my binoculars, I instantly knew it was a wild cat at that moment and asked my partner to look at it also, which she did, then as I went to look again it was there for a last fleeting look as I saw it snake through the marsh grass towards the woods. The adrenalin pumped round my body as the reality of what I had seen sunk in.

I have been brought up around cats all my life, and am fully aware of the size of domestic and even feral cats and this was neither. It was much larger (aprox 3 feet) and moved differently. I think it sensed I was observing it or smelt my scent which prompted it to disappear as quickly as it had appeared. The whole sighting lasted about 2 minutes.

Andy is a good friend of mine and has written about a Coypu in a Yorkshire attic in Animals and Men 19 page 20.

In December 2010 Marcus Matthews e-mailed me and said:

There have been two wildcats found- one at Swindon and one near Salisbury.

Also one near Milton Abbas woods.Wildcat reports do crop up sometimes- I have not come up with anything concrete. (1)

REFERENCE

1 E-mail from Marcus Matthews to Richard Muirhead December 8th 2010

A Scottish Wild Cat from the 1800s.Artist unknown. Reproduced with permission of Dr Karl Shuker.

A Spotted Cat from a 15th Century Hunting Manual

by Marco Masseti

Around about 1997 I bought a post card at the Bodleian Library,Oxford, of an intriguing spotted cat with a long,thin,tail (see illustration on back cover) which at the time I couldn`t identify (see also Animals and Men # 16 1998-1). Fortunately in April 2010 Marco Masseti provided me with enough information about it to find out much more about it. The image is from *The Master of Game*, translated into English by Edward of Norwich,2nd Duke of York. The Bodleian class mark is MS Bodley 546,fol 40 verso.This is what Masseti had to say:

Dear Richard,the image from `The Master of Game` you sent me represents a felid of medium size,characterised by a spotted coat. But,it does not represent any European or African wild cats of the *Felis silvestris* Schreber, 1775,taxonomic group.These are in fact characterised by striped - and not spotted - coats.

The representation of "wild" cats with striped and spotted coats is also found in the "Livre de chasse" written by Gaston Phoebus between the years 1387 and 1389 (see attached file;cf.De Urquijo et al., 1994; see also Longevialle & d`Anthenaise, 2002).But, I think that the latter spotted cats are feral animals,since spotted coats are not known in the wild European population. And,in any case,the spotted cats Gaston Phoebus are very far from the coat patterns of the cat of `The Master of Game` which features black spots on much paler ground colour of the coat.

Medium-sized cats characterised-as the felid represented in `The Master of Game` - by a long tail and a spotted coat are today found in Africa- the small black footed cat, *Felis negripes* Burchell, 1824 - and in the Middle East and south-eastern Asia, such as the marbled cat, *Pardofelis marmorata* (Maertin, 1837), the leopard cat, *Prionailurus bengalensis* (Kerr, 1792), the flat-headed cat, *Prionailurus planiceps* (Vigors & Horsfield, 1827), and the fishing cat, *Prionailurus viverrrinus* (Bennett, 1833).But I don`t believe that the image from `The Master of Game` has to be referred to one of the latter species.We have also to exclude all the American spotted cats due to the fact that America,at that time, was still far to be discovered.In medieval Italy,spotted cats and other wild spotted felids of the genera *Lynx,Caracal* and *Leptailurus,* were ordinarily referred to as *gattopardi*.

In my opinion,in the case of the spotted felid of 'The Master of Game' we are dealing with a non-veristic artistic representation.this is not a real cat but the evocation of a *leopard cat,* or-better-of its "idea." The morphological rendering of the animal is too much approximate and not accurate at all,and we can believe that the painter was not familiar with the subject portrayed. He certainly did not have used a live specimen as a model. Thus, this spotted carnivore might be generally referred to a felid species of unknown identification.but of the size of an European wild cat! Furthermore,it can be assumed that,rather than the portrait of a biological element known to the artist, the spotted felid represents a free elaboration of an iconographical model.Whatever its source of inspiration,the image seems to portray a species completely unknown to the artist.

REFERENCES

Cummins J. 1988 - The hound and the hawk. The art of medieval hunting. Weidenfeld and Nicholson,London;306 pp

De Urquijo A Thomas, M & Avril F (eds) , 1994 - Gaston Phoebus. El Libro De La Caza. Editorial Casariego, Madrid: 452 pp

Longevialle C de & Anthenaise C., 2002 - The Hunting Book of Gaston Phebus. Bibliotheque de l'image, Paris/H.Kilczkowski-Onlybook, Madrid: 96pp

Masseti M., 2009-Pictorial evidence from medieval Italy of cheetahs and caracels,and their use in hunting,*Archives of natural history*,36 (1) :37-47

Zebro - an Equine Mystery from Iberia

by Karl Shuker

During the Middle Ages and the Renaissance,several Spanish hunting treatises alluded to a mysterious,now vanished equine creature known as the zebro (or encebro, in Aragon), living wild in the Iberian Peninsula. In one of these works, it was described as "an animal resembling a mare, of grey colour with a black band running along the spine and a dark muzzle." Others likened it to a donkey but louder, stronger, and much faster, with a notable temper, and whose hair was streaked with grey and white on its back and legs. What could it have been?

Although largely forgotten nowadays, the zebro experienced a brief revival of interest from science in 1992. That was when archaeologists Carlos Nores and Corina von Lettow-Vorbeck Liesau published a very thought-provoking article in the Spanish scientific magazine Archaeofauna, in which they boldly proposed that the zebro may have been one and the same as an equally enigmatic fossil species – Equus hydruntinus, the European wild ass.

The precise taxonomic affinities of this latter equid have yet to be satisfactorily resolved, for although genetic and morphological analyses suggest that it was very closely related to the onager E.hemionus, one of several species of Asiatic wild ass, it can apparently be differentiated from these and also from African wild asses by way of its distinctive molars and its relatively short nares (nasal passages). Arising during the mid-Pleistocene epoch, approximately 300,000 years Before Present, the European wild ass persisted into the early Holocene before finally becoming extinct. During the late Pleistocene, its zoogeographical distribution in western Eurasia stretched from Iran in the Middle East into much Europe, reaching as far north as Germany, and it was particularly abundant along the Mediterranean, with fossil remains having been recovered from Turkey, Sicily, Spain, Portugal, and France.

According to Nores and Liesau,moreover,this species may have survived in southernmost Spain and certain remote parts of Portugal until as late as the 16th Century (they consider its disappearance to represent the Iberian Peninsula`s last megafaunal extinction), where, they suggest,it became known locally as the zebro. More recently,their theory gained support from the discovery of E.Hydruntinus remains at Cerro de la Virgen,Granada, dating from as late as the 9th Century.

Some researchers have also suggested that before dying out,the zebro gave rise at least in part to a primitive, nowadays endangered Iberian breed of donkey-like domestic horse called the sorraia (which was once itself referred to as the zebro.)

Furthermore, many believe that it was from the term ` zebro` that ` zebra` originated as the almost universally-used common name for Africa`s familiar striped equids

Even today,many Iberian place-names still exist in which the mysterious but now-obscure zebro`s name is preserved. These include Ribeira do Zebro in Portugal; and Valdencebro (in Teruel), Cebreros (Avila,) Encebras (Alicante), (Murcia) in Spain.

FURTHER READING:

NORES,Carlos & LIESAU,Corina (1992). La zoologia historica como complemento de la arquezoologia. El caso del zebro. Archaeofauna,vol. 1, pp. 61-71

MAD SCIENCE AND NATURAL HISTORY, VICTORIAN STYLE

HUNT EMERSON

www.steampunkandphenomena.com

A ZEBRA-DRAWN CARRIAGE

by Richard Muirhead

The image below, which was kindly provided by the Cheetham Library in Manchester, shows a zebra drawn carriage of the Mazawattee tea company.

According to Wikipedia: "The Mazawattee Tea Company was one of the most important and most advertised tea firms in England for around 50 years. Traditionally the origin of tea-drinking lies in China and the famous Tea Clipper ships raced across the seas to bring tea to London. In the eighteenth century, tea had become an important drink in Britain especially for the wealthy, but it was not until the 1850s (by which time tea plantations had been successfully established in India, especially in Assam, and in Ceylon) that a real expansion occurred. The Densham family were at the forefront of this period of growth. Originally from Plymouth, they moved to London and managed to amass a fortune from the business in quite a short time.

The Denshams later owned fine properties in both Purley and Croydon and one of the founder's sons, Edward, became a well-known figure in Purley..............The death of John Boon Densham at the age of 72 at his home in Croydon in 1886 ended the first period of the firm's growth. John Lane Densham was immediately made a partner and tackled the problem of the firm.

He decided to supply its tea in packets to retailers and in a different way by inventing a name for the firm. Being a great advocate of advertising, he reckoned that something quite unusual might be the answer and went to the Guildhall Library to get ideas. He came up with the idea of using the word "Mazathawattee, " perhaps based on the Hindi "Mazaa", which means " pleasure or fun" , and the Sinhalese "vatta", which means " a garden". This was shortened to "Mazawatte" and duly registered as a trade mark for retail sales in May 1887." (1)

The company effectively came to an end in 1953.

REFERENCE

1. Wikipedia – Mazawattee Tea Company
http://en.wikipedia.org/wiki/Mazawattee_Tea_Company

Two Notes on the Nandi Bear

by Richard Muirhead

Within the last few months I have found the two newspaper articles below which give interesting information about the Nandi Bear of East Africa. The first item is from `The **Singapore Free Press and Mercantile Advertiser**` of 28th December 1927.`

Nandi "Bear" may be a Giant Hyena.

Who will solve the mystery of the Nandi "bear", that strange beast of the East African veld which terrorised the natives and taken heavy toll of their stock and which devours its prey in a manner unlike that of any other animal? Asks the Morning Post. The Nandi "bear" has also taken human life, killing a child of the Karasai tribe and, since the tragedy, a number of people claim to have seen the animal.

The animal appears to bear a charmed life, for on every occasion when it has been sighted some unusual circumstance saved it from identification – and a Latin name. A rifle jammed, or the ammunition was finished, or an elephant was seen down stream.

Captain Ritchie, a noted big-game hunter and the Game Warden of Kenya Colony, believes in the Nandi "bear", and thinks it may be a giant hyena, though he admits " it may be different from anything we know". But why a giant hyena should have six digits, as the spoor of the Nandi "bear" suggests, he is at a loss to explain.

It is now clear that the perpetrator of a third series of hitherto unexplained crimes is a hyena of huge size. This beast, which recently killed twelve cattle near Tuso, does his work in a most unprofessional manner. In every case the carcase is found almost unscratched except for one shoulder, the near-by ribs and heart. The Kikuyu say that one of these redoubtable animals was killed at Tuso four years ago, that it was as big as a lion, and had ten spears through it before it died.

An expert who investigated the matter has no doubt that the animal is similar to the Strandwolf (Hyaena Brunnea) of South Africa, a specimen of which he obtained many years ago in the Kalahari, and which he describes as being very large, of a dark brown colour with darker spots and hair five inches long, hanging down the flanks.(1)

The second item below is exciting and suggests that there may somewhere be a pelt of the Nandi "bear" but the Natural History Museum in London couldn't help me with solving the problem of its current location (see below.)

This report is from 'The Hong Kong Telegraph' of 10th December 1936

Settler Shoots Beast of Legend.

White Settler Jesse R Coope, hunting in the Mau forest, 100 miles from here, has shot a huge lynx-like creature which local people here believe to be a Nandi bear. For more than 20 years natives and others have reported seeing a Nandi bear at rare intervals, but never has one been shot or caught before. The animal has achieved an almost legendary reputation.
Captain A.T.A. Ritchie, the Kenya game warden, says the shot beast resembles an outsize lynx but possesses significant points of difference. It has dark mahogany coloured fur. Local experts say they have never seen anything like it before.
The skin and skull are being sent to the British Museum for examination and possible identification. (2)

So on December 3rd 2010 I wrote to the Natural History Museum in London quoting the Hong Kong Telegraph report above and received the following reply later the same day:

Dear Mr Muirhead,

Thank you for your enquiry. Unfortunately I have not been able to find any mention of a J R Cooper or Nandi Bear in our Museum Archives.This does not necessarily disprove that it was sent here for identification-we have log books recording acquisitions,outgoing loans,and specimen swaps,but we do not have a consistent log of every specimen which was examined by our staff.

I have forwarded your email on to our mammals department in case they have any further information, but I'm afraid at present it appears we have drawn a blank on this.

If you have any further questions, please let me know.

Thanks,

Daisy.(3)

On December 6th I replied:

Dear Daisy

Thanks for your reply but the name was J R Coope, not Cooper, would that make any difference? Also a Captain A.T.A. Ritchie, a game warden in Kenya was involved.

Does your mammals department have access to your store room and could the skin have been moved during the War?

Yours sincerely, Richard Muirhead (4)

To which Daisy answered:

Apologies for my mistake, although I'm afraid the name Coope has not thrown any greater light on your query, and we don't have anything for A.T.A. Ritchie either.

Just to clarify on your storage query, we here in the Archives are only responsible for managing the paper records of our acquisitions and our institution, it is the Mammals Department who actually manage the specimens themselves. They do not have access to our store room, but our store room would not have been where the specimen would have been stored if we had had it.

Some specimens we held were temporarily moved to other storage during the Second World War. It's hard to say whether this would have been one of them until we've ascertained whether we held it, and if so when.

Thanks,

Daisy (5)

On December 7th I persisted:

Thanks Daisy, is there any way you can pursue this further? Also, does anyone in your Mammals department know any experts on Kenyan fauna, particularly of the Mau Forest? (6)   Daisy replied on the same day:

As I mentioned, I have forwarded your enquiry to the Mammals Department for them to follow up, they may take a little time to research into your request, but if we don't hear back from them when they've had sufficient time to do some digging, then I am happy to follow up with them again.

I have forwarded your enquiry regarding Kenyan fauna to our Botany Library, they should be in touch with you presently.

Thanks

Daisy (7)

This was the last I heard about the missing Nandi Bear skin and pelt. If anyone can provide Flying Snake with definite evidence of its fate or current whereabouts, please contact me and you will receive the next 7 issues of Flying Snake free and a complimentary copy of the ground breaking album Q: Are We Not Men? A: We Are Devo! . First come, first served!

REFERENCES

1   The Singapore Free Press and Mercantile Advertiser 28th December 1927
2. The Hong Kong Telegraph 10th December 1936
3. E-mail from Daisy to Richard Muirhead December 3rd 2010
4. E-mail from R.Muirhaed to Daisy December 3rd 2010
5. E-mail from Daisy to R.Muirhead December 7th 2010
6. E-mail from R.Muirhead to Daiisy December 7th 2010
7. E-mail from Daisy to R.Muirhead December 7th 2010

The Director and Management of the CFZ wish everyone at *The Flying Snake* success with their new venture

The Centre for Fortean Zoology, Myrtle Cottage, Woolfardisworthy, Bideford, North Devon EX39 5QR

Telephone 01237 431413 Fax+44 (0)7006-074-925
eMail jon@eclipse.co.uk
www.cfz.org.uk

54

Notes & Queries

Richard Freeman would like to know: Have there been any sightings of the Flying Snake of Namibia in the last 20-30 years?

Richard Muirhead replies: I typed in Flying Snake of Namibia into Google Books and came up with a reference to the South West Africa Annual of 1980. On pages 129-139 there is an essay by L.O.Honeyborne titled The Flying Snake. Honeyborne was a police sergeant in Keetmanshoop, Namibia, close to where a flying snake was seen in 1942.

Richard Muirhead would like to know: There is a mention in the manuscript of `The Inns of Salisbury` by R.G. Gordon of animal that was either a hyena or a "camel". The text of the manuscript describes the visit of a visit of a man to an inn called the Maidenhead in Salisbury. He wrote, on September 28th 1767:

The stupendous Hyena is a perfect Phenomenon: this astonishing animal can exist many days without either solids or fluids, and the latter more than 20, though in the most sultry climates, weighs upwards of 30 cwt, and is about 20 hands high" (? Camel.) – which " intends visiting this city", but does not seem to have done so.

I wrote to the late Clinton Keeling to ask his opinion as to the nature of this animal and on May 18th 1996 he replied:

" I was most intrigued by the strange "Hyena" of 1767, and am inclined to agree with you it sounds uncommonly like a Camel. In fact, Camels were not unknown in the country to date, as c.1610 several had been presented ex Spain, to King James 1; it`s more than possible, though, that none had been seen in the country since and people had simply forgotten what they looked like. On the other hand, there was then plenty of trade with North Africa, Turkey, India, etc., where they would have been seen by many Englishmen. The mystery deepens when it`s borne in mind that Hyenas, almost from the word "go", were common in travelling shows, and certainly the Striped species was exhibited at the Tower menagerie in the 18th Century. "

Can anyone else throw light on this matter?

The November 2009 a letter from David Green appeared in **BBC Wildlife Magazine**, page 111, as follows:

HERE BE WALLABIES

"Elisabeth Wallace`s lovely story about a red-necked wallaby on the loose in Devon (Tales from the bush,September) brought back memories of my father,who used to see a group of pure white wallabies, as well as normal coloured ones, on the Derbyshire moors before World War II. I remember seeing them myself in the 50s, as a child of 10, but unfortunately I never spotted the white individuals. "

David Green
Derbyshire

Richard Muirhead would like to know: Has anyone any more information about these white wallabies? "**Richard Muirhead** would also like to know: Does anyone know of any cryptids in Cuba?

Dr Devo about 2 years old.

Copyright The Editor

56

In the Beginning was the End Oscar Kiss Maerth

This extraordinary and worryingly eccentric or subversive book, depending upon your point or view, or degree of sanity/insanity, is concerned with, as the blurb on the back cover puts it:

" A Shattering New Theory of Evolution! This startling bestseller poses a devastating new theory of evolution that explodes the accepted that explodes the accepted concepts of Darwin! It claims that Man evolved through the cannibalistic practices of primeval apes! That intelligence can be eaten. That the cannibalism of his ancestors still preys upon the conscience of Man and has caused him to lose his innate powers of extra-sensory perceptions!"

It has been said that Oscar Kiss Maerth is a pun on Oscar Kiss My Ass. Well this piece of anthropology may have impressed Devo around the mid to late 1970s, but around 35 years later it fails to move me although I am a Devo fan. For example to quote from chapter 8,

" Genesis is an intuitive description of the beginnings of life on earth, notably of the abnormal development of a hairy animal into man; who by eating the fruit of knowledge, has become naked, sexually disordered and intelligent. As a result of this unnaturally acquired intelligence mental disorder developed and with it man's delusions, burdening him with the diseased concept of work and progress" (p. 203)

Now it may seem a minor point but the tree in Eden was a tree of *the knowledge of good and evil* not simply knowledge. There is nothing wrong with knowledge, the case was that Adam and Eve were not ready for knowledge at that moment in time. This knowledge brought about spiritual death, which is probably one of the ways cannibalism came into the world in the first place.

Maerth says:

" One ape discovered that eating the fresh brain of one's own kind increases the sexual impulses. He and his descendants became addicted to brains and hunted for them. It was not until later that they noticed that their intelligence increased as a result. The outcome of this process is HOMO SAPIENS (p. 37)

It is hard to know what to make of a book like this. It is daft. Dangerous even. However it is also a brave attempt to challenge evolutionary thinking and religious dogma. Let Devo have the last words:

" The beginning was the end, of everything now, the ape regards his tail, stuck on it! Repeats until he fails, half a goon and half a god...." (Devo - Gates of Steel)

Moa Sightings Bruce Spittle Dunedin New Zealand: Paua Press Ltd 2010
Vol 1 (ISBN 978-0-473-15356-4) 2 (978-0-473-15357-1) 3 (978-0-473-15358-8)

I knew next to nothing about the evidence for the survival of the moa in New Zealand from c. 1150 (chapter 110) up to the (in?) famous 1993 Craigieburn Forest Park sighting (chapter 151.) However, I am now able to thoroughly recommend Bruce Spittle's three volume ` Moa Sightings` .But book a six month holiday on some remote unexplored island to fully absorb this thoroughly researched body of work. This is not a book for casual reading but a serious reference source.

The caption on the back cover of each volume explains:

" Hunting pressure, habitat destruction, and introduced predators led to moa extinction by A.D 1650 according to the previously held serialoverkill model. In the currently accepted rapid "blitzkrieg" model, all the moas were gone by A.D. 1450, over 300 years before the first Europeans landed with Captain Cook in 1769. However, a number of moa sighting claims have been made since 1769 and the author offers for consideration a staggered survival model in which moas lingered on until a later date in some remote, isolated areas. The available circumstantial evidence for a few moas remaining after 1769 is presented including reports suggesting survival in circa 1810 by Kawana Paipai, 1845 by Burr Osborn, 1863 by Patrick Caples, circa 1825-1875 by HJ Cuttance, and 1878 by Sir George Grey.
The author, Bruce Spittle, is of part-Maori descent and lives in Dunedin, New Zealand.

The volumes are divided into chapters giving name(s) of the witnessing and the " claimed time and place" for the moa being alive. Taking volume 1 chapter 3 as an example:

John Boultbee 1826, Milford Haven (Milford Sound)
Introduction, The Claim.
Discussion.

The chapter is well illustrated with a map and photos, as are all the chapters of the volumes. Each volume is comprehensively indexed. One criticism I do have is that perhaps the chapters could have been arranged chronologically. However at NZ$70 a volume with free shipping these are expensive but very worthwhile.

Letters to Flying Snake

There have been no letters as yet to Flying Snake,because only a handful of people know it exists,so I thought I`d include a few letters I`ve received over the last 20 years or so which I hope you might find of interest. This first one was from Dr Anthony Bogadek then of St.Louis School,Hong Kong,dated 2nd March, 1997, joint author of Hong Kong Amphibians and Reptiles 2nd edition with Stephen J.Karsen and Michael Wai-neng Lau which was published in 1998.

Dear Dr (sic!) Muirhead

Thank you for your letter of 18-2-97.The way you spelt my name (Bovidec)would suggest that I belong to the family Bovidae! Don`t worry!It`s quite OK with me.People have mispelled and mispronounced by(sic) Slav name hundreds of times.Its great fun for me.

I am afraid I have no answers for any of your intriguing questions.

Your first question is a little puzzling.You are asking whether there are any snakes in Hong Kong "at present unknown to science." What might or might not be present here, no-one can tell until time as the species is discovered. In recent years, we have made a few new records of Hong Kong snakes, but these species are not new to science. They occur in nearby Guangdong Province.They include *Amphiesma boulengari* and *Rhabdophis nuchalis*. The only local reptile species,known to be new to science, are *Dibamus bogadeki* , afossorial lizard discovered a few years ago,and Hemiphyllodactylus sp., a new gecko species still unamed and under investigation.In our forthcoming second edition of "Hong Kong Amphibians and Reptiles", updated and very much expanded, there will be a short chapter on the history of herpetology in Hong Kong and a list of some 20 herp species of possible occurrence in Hong Kong.

Snakes reported from Hong Kong once before "but never seen again" include *Cylindrophis ruffus,* a specimen of which is now in the Museum of Natural History (Vienna,Austria); *Chrysopelea ornate*(its whereabouts are unknown) and *Dendrelaphis pictus.* A Dendrelaphis specimen was caught after a gap of 90 years at Shek Kwu Chau, a small island south of Hong Kong.On the same island were found two specimens of *Ahaetulla prasina,* a new record for Hong Kong.However, no one can tell whether these four species are part of our native fauna or were introduced into Hong Kong by the food and pet trade and later released,or whether they are escapees from captivity.

There have been past reports of elephants living in the wild in southern China (see pamphlet by A.A Fauvel on "The Alligator in China", published privately by the North-China Branch Royal Asiatic Societ Shanghai (1879).Whether elephants as well as other species mentioned by Fauvel ever strayed into Hong Kong is as good a guess as any.

I am not aware of any accounts, Chinese or otherwise , specifically on the natural history of Hong Kong. I would be happy to have the information should you ever come across any sources during the course of your work. And,finally,no one of my acquaintances knows what has happened to the museum collection that you refer to. I was very interested in tracking it down because Chrysopelea was in that collection. Because we could not locate it, and in the absence of further evidence, we were compelled to discount this snake as a valid member of our Hong Kong snake fauna.

I am sorry that I have not been very helpful. You may find Er-mi Zhao and Kraig Adler`s " Herpetology of China", published by the Society of Ambhibians and Reptiles (SSAR); ISBN: 0-916984-28-1, a useful source of herpetological references.

With my best wishes for the success of your research and my best personal regards.

Sincerely yours

Anthony Bogadek

In January 1998 I corresponded with the late Clinton Keeling on the subject of melanism in animals:

16th January 1998

Dear Richard,

Greetings..............Regarding melanistic Squirrels, when this sort of situation occurs in an isolated or localised area (and it need not necessarily concern colour;for example,it`s exactly the same with,say,the Black Rhinoceroses in one part of Kenya which have no external ears)it`s usually because a particularly dominant male or fertile female has been carrying what might be termed "baddie" genes. The same applies to the black wild Rabbits of Anglesey and the Water Voles of Norfolk. For what its worth,though, one observation of my own is that very frequently, although by no means invariably,animals in districts where the soil is acid are often darker than their counterparts where it`s more alkaline or limey.

Where albino Squirrels are concerned, I know of no British populations or concentrations, BUT the Red is far more likely to throw such individuals that the Grey-in fact when the former was commoner than is now the case most of the white ones seen were of that species. Why,though,Isn`t immediately clear.

Hope this is of some use to you.

Kind regards,Clinton

A missionary in the Republic of Congo Congo (as it then was) replied to a query of mine in June 1995:

6-6-95

Dear Richard

I am a missionary with United World Mission doing church planting work with the Aka in the Republic of Congo.My family and I arrived back on the field in Jan.1995.We will be here for 4 years.UWM

I personally believe that there may have been an unknown animal in our inundated forests, but it may now be extinct.One of my colleagues,Gene Thomas,strongly believes that it exists. You can reach him at BP 24 Brazzaville Republic of Congo Phone & FAX 00-242-82-33-07

The testimony of the Aka and the Bantu villagers is not conclusive. They can tell you about the dinosaur, but they can also tell you about men who live in the water and gorillas who talk and write letters. The only way to separate scientific fact from legend is if someone got some hard evidene.

The 80's were the heyday of the Mokele Mbembe expeditions. Each one that went to Lac Tele "saw it" but they weren't able to get it on film.In the early 90's large scale scientific expeditions using underwater photography found nothing(The Japanese filmed a TV show of an expedition.They left a big debt with the villagers so I pity the next expedition to Lac Telle) There is a lot of wildlife research going on in the Congo so maybe someone will see the thing. Who knows when God might choose to reveal His secrets.

And now to that hero of modern British Socialism, Ken Livingstone MP for Brent East (1987-2001) as he then was. As one of the extremely small band of Macclesfield socialists, I present to you,comrades, Ken Livingstone on the Eft!

22 April 1996

Dear Richard Muirhead

Thank you for your letter of 15 April about the terms eft

Or evvet. They were just other names for newts - not crocodiles or anything else.

Yours sincerely

Ken Livingstone.

Wikipedia Creative Commons

The following letter is included in honour of my late father Stuart W.Muirhead (1931-1993) as it includes some items of interest from Hong Kong,which is where he nurtured my interest in zoology. The letter is from Ho Shai Lai, a former General in the Kuomintang, who fled China after the Communists took power in 1949.

April 18,1996

Dear Richard,

Thank you for your letter of March 27,1996.I have sent a copy of your letter to my sister,Irene.Her address is: Dr.Irene Cheng, 4730 Noyes St., # 311,San Diego, CA 92109 U.S.A.

Although she is well over her 90's I am sure she will be delighted to answer your questions. Regrettably I have not read her book on my mother,Lady Clara, as I am not a good reader. Furthermore I do not have much memories of 50 Peak Road,Mount Kellett, as I only stayed there for about five years before my father sent me down to Idlewild in care of my other mother, Lady Margaret.She had no children and since I was a "problem child", she had more time to train and encouraging me to live a respectable life. At the Peak, Lady Clara had nine children. Anyhow the following may be of some help in answering your questions:

1. I remember the huge tortoise very definitely. I used to stand on it and it and it would move around without any problem. Its back shell had a diameter of about 10 inches. It was brown with black spots or black stripes.To the best of my memory, I have never seen a tortoise of that size.Where it came from, I do not know.With the completion of a temple which Lady Clara had built in Happy Valley in the 1930`s the tortoise was moved over there. It lived at the Tung Lin Kwok Yuen,(the temple) for many more happy years after the end of the Japanese war.

2. There was some wildlife around our house at Mount Kellett. Among them were many poisonous cobra and bamboo snakes. There were also a few pythons which came to steal our chickens. One night a very long python, about 12 feet long was caught on our tennis court. I believe my father presented it to the Hong Kong Museum where it was placed on exhibition for many years. Since I have not been to the new museum, I do not know whether the python is still on exhibition.

3. There were also a number of "fruit foxes" that ate up the fruits on the trees and the vegetables from the garden. One night we engaged a caterer to serve dinner for a family party, and he noticed a "fruit fox" perched on a tree in the garden. He tried to catch the fox but the fox bit him so badly that he could not serve the dinner. Our dogs always tried to kill these foxes. They succeeded

In doing so once or twice a year but nowadays the Peak is so heavily inhabited and I haven`t seen a live "fruit fox" for many years, except the remains of one that was killed by our shepherd dog two years ago. This is all the information I can furnish you as I said I only stayed on the Peak for about five years.It was not until 1961 that I moved back with my family.

Your father visited us soon after we moved to 75 Peak Road. If you or your mother come to the Far East and if we still live at 75 Peak Road, you will be most welcome.

Your father gave us a lot of good advice and took care of my father`s estate.We deeply appreciated it. As I was a military man when my father died we did not know how to manage the estate. We still run it according to most of the advices of your father and so far we seem to be getting along fine. My children and my family remember him with respect. He provided genuine beyond that of a normal trustee. I went to his memorial service as I wanted to show my respect and appreciation....

Yours sincerely,

Ho Shai Lai

AN **ORANGE-COLOURED BADGER** IN MATLOCK , DERBYSHIRE UK, 2005

This photograph was taken at the One World Festival in New Mills, Derbyshire on July 3rd 2010.

The badger was found dead by the side of the road in Matlock, Derbyshire, 2005.

SPOTTED CAT from The Master of Game - A Fifteenth Century book in the Bodleian Library Oxford (MS Bodley 546 · fol 40 verso). See Pages 44-45 of Flying Snake 1.

Flying Snake

A Journal of
Cryptozoology, Folklore and Forteana

Volume 1 Issue 2   **October 2011**   **£3**

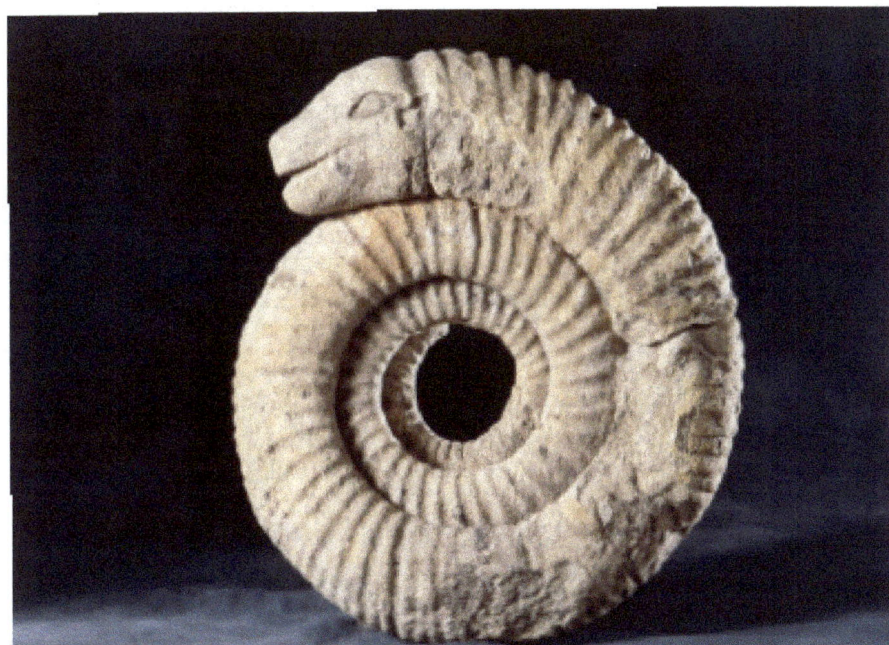

In This Issue: Pine Martens in Derbyshire, 1996-2011 •
Red Heifer and 3rd Temple • Spotted Otter in Ireland •
Lazarus Syndrome • Oddity in Forest of Dean • Weird
Worms • And More!

ABOUT FLYING SNAKE

Flying Snake is available from:
Richard Muirhead
Flying Snake Press,
112 High St,
Macclesfield,
Cheshire,
SK11 7QQ
UK
http://homepage.ntlworld.com/richmuirhead/cryptozoology/
Tel: 01625 869048
 Mike Hardcastle,Sub-Editor,NSW Australia .
www.steampunkandphenomena.com

My initial plan is to publish 3 issues of Flying Snake a year. Please feel free to contact me if you want to reproduce anything I have written. If you want to reproduce other authors' works, I will try and contact them on your behalf and get back to you. The opinions of authors other than myself do not necessarily reflect my own.

Back issue available on request.

A PDF version of Flying Snake is available for £3 per copy.

PAYMENT

Subscriptions: £ 3 per issue, £ 9 per annum.

Payment for however many issues you (and your friendly neighbourhood, reading, flying, snake) would like to purchase can be made by means of PayPal on my web site (See url above).Or via http://www.flyingsnakepress.co.uk

Cheques and postal orders from within the United Kingdom should be made out to Richard Muirhead, and *not* Flying Snake.
Cheques will not be accepted from outside of the U.K at this point in time.

CREDITS

Flying Snake Title/logo: William Frederiksen

Image on cover: Snake stone, converted fossil ammonite. From front cover of `Formed stones,` Folklore and Fossils by Michael G.Bassett. Reproduced with permission of National Museum of Wales.

CONTENTS

DR DEVOS DIARY

"For I pray God for the introduction of new creatures into this island. For I pray God for the ostriches of Salisbury Plain, the beavers of the Medway and silver fish of Thames." Christopher Smart, `Rejoice in the Lamb`. Poet, naturalist,lunatic (1722-1771)

Dr Devo has been busy in his Recombinant DNA Parlour since Flying Snake 1 was published in April, researching and requesting new cryptozoological and Fortean articles from the outer fringes of these disciplines and whilst trying to get out of his yellow plastic jump suit, has managed to put together the cocktail of essays you now have before you. I must admit my hyper-sensitive conscience has been troubled by the lack of " mainstream" cryptozoology in issues 1 and 2. However, I believe these areas are well covered elsewhere. By " mainstream" I mean Alien Big Cats and the Loch Ness Monster for example. Of course what is mainstream to me may not be the same to you.Lizzy Clancy has written about the Lazarus Syndrome so I am not altogether obsessed with the totally obscure: I hope you are Biblically up to scratch - this is not about the 1970s folk band Lazarus and I hope you will find much to interest you here.

Talking about the Loch Ness Monster, the other day I was told of a sighting in the early 1970s by someone I trust which I`m hoping she`ll allow me to write about in issue 3 so that's something to look forward to. I am also busy with my approximately twice a week `Muirhead's Mysteries` blog on Cryptozoology On-line which Jon Downes of the Centre for Fortean Zoology has decided to produce in book form in about two years or so. The Mystery Animals of Hong Kong book continues to gather new information like a Ropen gathers fish or even dead human bodies if some sources are to be believed. Meanwhile the non-cryptozoological part of my life continues,an endless round of karaoke,Oxfam,library visits, drinking Rubicon fizzy Lychee flavoured drink, recording the rainfall and other somewhat geekish activities. But enough about me,lets think about you for a while.......Please read on! And enjoy .

" It is the simple truth that man does differ from the brutes in kind and not in degree; and the proof of it is here: that it sounds like a truism to say that the most primitive man drew a picture of a monkey and that it sounds like a joke to say that the most intelligent monkey drew a picture of a man. Something of division and disproportion has appeared; and it is unique. Art is the signature of man." G.K.Chesterton Journalist,theologian (1874-1936)

A Strange Creature in the Forest of Dean, 1924

Richard Muirhead

I came across this story in the mid 1990s but dumped it because I thought it too weird!-Typical.Thanks to Dave Tuffley for permission to quote from the newspaper extract below. Thanks to Rob Wilkes and his father for drawing my attention to this story about 20 years ago.

Dean Forest Guardian 15th August 1924

Strange animal found in coal mine

An amazing amount of space was devoted by newspapers,although apparently overwhelmed with news of national importance (to wit) the budget, concerning a journalistic story sent from the Forest about a "living fossil." Miners working on the bank of the Poolway Level, near Coleford,discovered in the coal that they were handling for transport,what appeared to be the body of an animal 14 inches in height. The animal showed signs of life for about a hour. The object was left on the coal bank for about a night and the next day it had disappeared. On this raw material a monumental amount of conjecture has been superimposed and promptly demolished. South Wales Contemporary has been giving a half column about the discovery and went to the trouble of contacting one of the most distinguished naturalists in Wales. After he had finished laughing he said this "This is the answer - a bat! A bat at this time of year fills every condition except that of size." And the estimate given by our correspondent is admittedly that of nervous men afraid of a weird creature. A bat hibernating through the cold months seeks a cave or crevice and there hangs with head downwards with its wings so tightly pressed behind it that looking at it from the front or above as it fell on its back they could not have seen and would have given the impression of an armless trunk with well arched shoulders. Shaken from its rocky perch or crevice and fallen and would have partly awaken a month too early and would lie almost without movement and with no effect to escape.

We have made no further reference to the mater than above but for the fact that so reliable a witness as Mr Leslie Jones,the discoverer, remains so steady in his belief that the discovery was something out of the ordinary and certainly no bat. He is indignant that anyone should imply that he does not know a bat when he sees one. He claims to be well versed in fossils and being a well known ambulance man, nothing will move him from the fact that the creature had life.

It was on Saturday previous to Easter [April 20th -Richard] and working in the pit bank at Poolway Level owned by Mr Amos Brown of Wynols Hill, that the incident occurred. The previous day the men employed at the colliery had cut through a piece of coal into the original workings of what is now styled the New Hawkins Colliery, of which we believe there is no record at all at the Crown Office.

Mr Jones was using a shovel getting coal ready for transport when he saw a movement amongst the coal which had been taken from the old workings. He became interested enough to investigate. He immediately saw that the shovel had struck a small object, very much resembling a human being which he threw on top of the bank. Calling his father and Mr Amos Brown together with a passer-by named Tye, they proceeded to examine the object of attraction and were astounded to find it still living. The shovel had struck it on its hind-quarter. With the aid of sticks, they made a thorough examination with the exception that it possessed no arms, they declared that the creature was almost a perfect model of a human body and it was estimated to weigh nearly two pounds. It was from 12 to 14 inches long and had a round head the size of a teacup. With ears, eyes,nose,mouth and tongue and with teeth in both jaws. It was entirely covered with hair of a brown colour that on the face being slightly shorter than that on the body. The hair on the body was nearly an inch in length. The body was some 9 inches in circumference and the jointing of the legs and feet were perfect, even to toes and nails. The formation of the mouth was also perfect and the tongue was of a pink colour and rounded off. Although the creature was seen by several people at the time, no one thought it of any interest and the object was left on the bank when work finished for the day.

On Easter Sunday morning, Mr Jones was sufficiently concerned to visit the spot again only to find no evidence of the previous days find and continued searchings have failed to bring the object to light again. In most quarters the story is laughed at but not by the men who affirm that the remarkable little creature was more like a human being than anything else. It was thought that someone who had heard of the discovery, secured it but even now hope is entertained of its whereabouts.Mr Jones scoffs at the suggestion that it was a hibernating bat, referring to the fact he has yet to see a bat the size of a kitten, weighing in at nearly two pounds.

He concluded the interview with our representative by saying that the examination made was very thorough and the naturalist's opinion may have been arrived at by the reports that the head of the creature was about the size of a peacock, which was an understatement. (1)

Significantly, the creature was covered in hair and although compared to a bat, wings are not mentioned. Could it have been an aborted foetus? The largest bat it could have been would possibly be a Mouse-eared bat (Myotis myotis).Or the Greater Horseshoe bat (Rhinolophus ferrumequinum) . One could have strayed from near the South Coast of Britain to the Forest of Dean and these bats use caves. Poor lighting, shadows and imagination could have caused the observer to mistake several greater horseshoe bats in a tight cluster as one bat. Supposing it was a bat and that is a big supposition, could it have been an escaped fruit bat? Or perhaps it was an owl ? We will probably never know.

REFERENCE

1. Dean Forest Guardian August 15th 1924

Lazarus Syndrome

Lizzy Clancy

They took therefore the stone away. And Jesus lifting up his eyes, said: Father, I give thee thanks that thou hast heard me. And I knew that thou hearest me always: but because of the people who stand about have I said it, that they may believe that thou hast sent me. When he had said these things, he cried with a loud voice: Lazarus, come forth. And presently he that had been dead came forth, bound feet and hands with winding bands. And his face was bound about with a napkin. Jesus said to them: Loose him and let him go. (John 11: 41-45)

This is how the story of Lazarus's resurrection four days after his death is related in the Bible. This event is believed by many to be one of Christ's most important miracles, but is by no means a singular occurrence.

61-year old Daphne Banks was seen to be breathing just moments before her body was due to be refrigerated in the hospital morgue. Her own doctor had pronounced dead at her home in the early hours of New Year's Day 1996. The police had been called because the GP believed a post-mortem should be carried out ; hence Mrs Banks's transfer to the hospital. The lady's vicar seemed to agree with Mr Banks that a miracle had occurred. The apparent death of Mrs Banks had described as ` sudden' , which is a feature that cases of Lazarus Syndrome share in common.

It later transpired that after epilepsy and mobility problems caused Mrs Banks to be unable to go out very much, she became increasingly depressed and had taken an overdose. Just a week later Daphne Banks was able to go home, in near-enough perfect health, notwithstanding the health conditions she already had.

In 2009 twenty-three-year old Michael Wilkinson was rushed to hospital in Preston and pronounced dead of a heart condition only doctors to discover a pulse half an hour after fifteen minutes of attempts to resuscitate him had failed. Unfortunately,Mr Wilkinson's recovery did not last very long and he finally died two days later after being moved to the intensive care unit.

'Lazarus syndrome or autoresuscitation after failed cardiopulmonary resuscitation is the spontaneous return of circulation after failed attempts at resuscitation' according to Wikipedia. An article in the *Daily Mail* puts the worldwide total of cases at 38, as does the conditions entry in the Free Online Medical Dictionary, though other sources put the total at variously twenty-four and twenty-five.

Dr Bruce Ben-David, in his online article **Survival After Failed Intraoperative Resuscitation: A Case of " Lazarus Syndrome "** suggests the possibility that a large number of cases exist but go unreported because of health professionals' fear of accusations of negligence or exaggeration. He also notes that it may be possible that such bizarre cases cause surgeons to doubt their own observations of the patient in question. Dr Ben-David also draws on previous studies to suggest a variety of potential causes of Lazarus Syndrome, including that the body may, for whatever reason, delay sending adrenalin to the heart when a patient goes into arrest, so that when it eventually does arrive, it kick-starts the heart after doctors have thought the patient already dead. He cites a colleague who advises a ten-minute period after apparent death for observation in case a patient returns to health.

But what about Lazarus himself? Of the cases I have discovered in my research I have yet to find a recent example of apparent resurrection where the formerly-deceased came back to life after much more than several minutes. Every one of them appears to have had their miracle within a day at the very most. Both Lazarus and Christ were supposed to have been raised four and three days respectively after their internment.

Rodney Davies suggests in The Lazarus Syndrome that Lazarus may not have actually been dead but in a cataleptic state. He cites Jesus's apparent nonchalence at the news that his friend was gravely ill and Christ's assertion before setting off for Bethany that Lazarus was only sleeping.Mr Davies also makes the point that often those in a ` cataleptic` trance can be brought out of it if shouted at and Jesus is reported to have 'cried with a loud voice' before Lazarus came forth. Davies says that decomposition hadn't set in with Lazarus's body and that this is further evidence of trance or coma rather than true death. However, I am not aware of any documentation of Lazarus's body being in perfect condition upon his recovery, so it would seem that Davies is assuming there was no putrefaction based on the idea that Lazarus had not been dead.

Also, Jewish custom at the time of Christ actually dictated that the tomb was re-opened after three days in all cases just in case the `deceased` was actually not deceased after all. Upon determining that the person really was gone to the hereafter, perfume and oil were applied and the tomb was re-sealed for another twelve months. It would not make sense for Martha to advise Jesus that her brother's body `stinketh` if she had not found that to be the case the day before when she and her family had opened the tomb.

In the case of Jesus himself, this custom explains the women's return to the tomb on the Sunday morning, though the traditional explanation is that the onset of the Sabbath prevented the full burial rituals being performed. And from a non-Christian perspective, it could go some way to explaining the resurrection story as an early case of Lazarus Syndrome, though many argue that the suffering from scourging and crucifixion could not allow for this possibility.

Whether you believe in miracles or not, Lazarus Syndrome, while being widely accepted as a genuine medical condition in mainstream science, has been a Godsend of some description for those who thought they had lost their loved ones only to have them delivered safely to them again.

 Bibliography

ANCIENT TOMBS:ARCHAEOLOGY,DEATH AND THE BIBLE,URL:

http://www.bible-archaeology.info/tombs.htm [16/09/2011]

Davies, R(1999) The Lazarus Syndrome. Burial Alive and Other Horrors of the Undead London: Robert Hale Ltd.

LAZARUS SYNDROME- DEFINITION OF LAZARUS SYNDROME IN THE MEDICAL DICTIONARY BY THE FREE ONLINE MEDICAL DICTIONARY, THESAURUS AND ENCYCLOPEDIA, URL:

http://medical-dictionary.thefreedictionary.com/Lazarus+Syndrome[16/09/2011]

LAZARUS SYNDROME MAN PRONOUNCED DEAD COMES BACK TO LIFE FOR TWO DAYS – MAIL ONLINE, URL:

http://www.dailymail.co.uk/news/article-1192283/Lazarus-syndrome-man-pronounced-dead-comes-life-days.html[12/09/2011]

LAZARUS SYNDROME – WIKIPEDIA, THE FREE ENCYCLOPEDIA, URL:

http://en.wikipedia.org/wiki/Lazarus_syndrome[12/09/2011]

SURVIVAL AFTER FAILED INTRAOPERATIVE RESUSCITATION: A CASE OF " LAZARUS SYNDROME" URL:

http://www.anathesia-analgesia.org/content/92/3/690.full?Ijkey=5942b3601eb043bcbdd82e37cb0d889a2b3a0625&keytype2=tf_ipsecsha [12/09/2011]

SWOON HYPOTHESIS-WIKIPEDIA,THE FREE ENCYCLOPEDIA, URL:

http://en.wikipedia.org/wiki/Swoon_hypothesis [16/09/2011]

WOMAN FOUND ALIVE IN HOSPITAL MORGUE-NEWS-THE INDEPENDENT, URL:

http://www.independent.co.uk/news/woman-found-alive-in-hospital-morgue-1322541.html [12/09/2011]

References in bold as in original manuscript

飞碟探索

4
1984

`U.F.O. Discovery`.Translation by Mr Shanshun Li from Superherbs Macclesfield,U.K.

FLYING SNAKES
PART TWO

Dale Drinnon

This article is part two of Dale Drinnon's original blog which first appeared in Cryptozoology Online January 23rd 2011

" I hypothesize that at the beginning of the post-glacial(recent) period the ancestral long-necked pheasants spread throughout Europe,starting in the southeast but then following the advance of the forests northward until they were spread over most of the area. They were rarer after the establishment of farming and retreated to the wilderness areas. By the time of the Roman Empire, they had been extirpated around the Mediterranean and the common (ring-necked) pheasants began to be imported from Russia (Scythia) to replace them[indicated by darkest grey on the map]By the Dark Ages they had disappeared in Central Europe and whereas before there were several different colour schemes for the males, in the later Middle Ages and on there were commonly on the green phase in Western Europe and the red phase in Eastern Europe. They would have died out much sooner except they had become associated with a superstitious dread that made people keep away from them. Reports became spotty in recent years but they were seen sporadically in the 1800s and possibly early 1900s. in England and Wales on the one hand, and in the Baltic countries on the other.

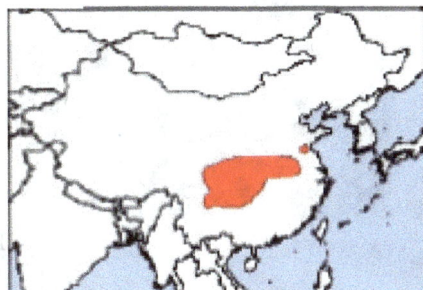

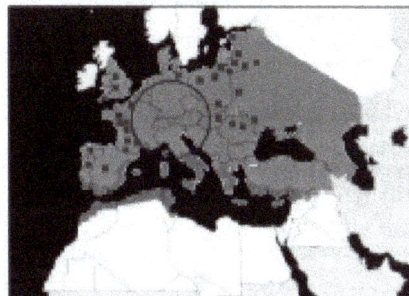

I have a contemporary report of " Venomous Flying Snakes" from Novgorod in Russia at the Yahoo group Frontiers-of-Zoology, and I now believe that report to belong in this category. Recent reports also seem to come from the Basque territories in Southern France and Northern Spain."

The images below and on the previous page appeared in Dale's original blog.

THE

Flying Serpent,

OR,

Strange News out of

E S S E X

BEING

A true Relation of a Monstrous Serpent which hath divers times been seen at a Parish called Henham-on-the-Mount within four miles of Saffron Walden.

Showing the length, proportion, and bigness of the Serpent, the place where it commonly lurks and what means hath been used to kill it.

Also a discourse of other Serpents and particularly of a Cockatrice killed at Saffron Walden.

The truth of this Relation of the Serpent is attested.
Richard Jeaffin – – Church-Warden.
Thomas Prophett – Constable.
John Knight – – Overseer for the Poor.

By
Barnaby Thorgood,
Samuel Garrod,
Richard Early,
William Green,
Householders.

WITH ALLOWANCE.

LONDON, Printed and Sold by Peter Lillicrap in Gutteridge Ally. [1669]

SAFFRON WALDEN;
Reproduced in fac-simile by W. Maitland,
With Introduction by Henry Miller Chitty.
1884.
Price Sixpence.

There were six comments;

Retrieverman said....

Does the CFZ still have the pair of Reeve's pheasants?

Ego Ronanus said...

Unfortunately, bits of this interesting article have been cut at the edges. Could a re-edited version be managed, as it certainly must rank as a diligent piece of research.

Dr Karl Shuker said...

Although intriguing, Dale's suggestion that the Welsh feathered snakes were cock pheasants is far from new. Way back in 1995, within my book 'Dragons: A Natural History', I noted: " It has been suggested that brightly coloured serpents with feathered wings spied in the vale of Edeyrnion in 1812 may have been cock pheasants , which were unfamiliar there.". However, I still find it difficult to believe that a pheasant could be mistaken for a flying snake - unless the latter term had a much wider meaning, such as 'flying dragon'.

Oll Lewis said...

The Gwibers possibly being pheasants or similar birds rather than snakes has always been one of my prefered explanations, like with all widespread cryptozoological reports though there are likely to be a variety of explanations, different in each case. For example Ned of Glamorgan forged a lot of folklore generally regarded as authentic by many even to this day among these were some tales of Gwibers

We have a male Reeves pheasant still at the CFZ

In the last picture Dale used it is interesting to note, how similar the firebird looks to the Chinese Fenghuang which is also based on a pheasant.

Dale Drinnon said....

Thank you all for your comments. I was unaware that a pheasant-like bird had been suggested as a solution to this mystery but this is also an explanation I have had for quite some time and only just now elaborated upon. Indeed my oldest notes with this suggestion as regards Wyverns and Cockatrices go back to the 1970s.

In specific reply to Karl`s remarks, the Anhinga is frequently spoken of as " A Flying Snake" by inexperienced observers and that is mainly ONLY for the long neck. I had thought to make the neck longer on my reconstruction but on the balance with the depictions I had on hand, I made it somewhat shorter in my reconstruction than I would have liked. The "Serpent" part comes mainly from the very long tail and the flight profile which makes it look as if it were all one elongated mass with wings added. And the part about both the head and the tail "coiling up" is explained in the entry: one of the photos shows the pheasant's tail " Coiling"

This is now my prefered explanatin for the Western, European Flying dragons and I prefer it over Pterosaurs and the like. The fourlegged kinds are a much more recent addition and Water dragons are are a separate category again.

Best Wishes, Dale D.

AnnF said...

Very clever idea, whether new or not! Ring-necked, at least, can move explosively quickly, so much that it could seem like a snake striking. If other birds have the habit of lying low and then bursting out unexpectedly, this snake-like behaviour would be more " evidence" that this thing is a snake.

FIGURES

Top left to bottom right going anti-clockwise. Page 13

1. Reeve's Pheasant tail curling up Snakelike

2. Reeve's Pheasant Original Range (below the pheasant)

3. `Flying Viper` Pheasant Hypothetical Original range, Biome similar to Reeve's Pheasant in China. Showing Eradications in Central area in early Historical period and Persisting in the Fringelands

4-7 Top left to bottom left going clockwise: Page 14.

4. " Flying Serpent" Longnecked Pheasant Mockup for Possible

Appearance in Life

5. "Flying Serpent" Pheasant to Scale to Human, Size as Commonly

Reported

6. "Guivre" Image

7. The Flying Serpent or Strange News out of Essex

The Quest for The Red Heifer

Zvi Ron

One of the more esoteric rituals described in the Bible is that of the red heifer (Num.19) .The ashes produced from burning a red heifer were used as part of the purification process for people who came into contact with a dead body. Paradoxically, those who burn the cow and collect the ashes themselves contract a low level of impurity. The ashes were mixed with water and sprinkled on the impure person on the third and seventh day of the purification process. Purity was considered very important in ancient times, since it is prohibited to enter the Temple in a state of impurity. Even in modern times, according to Jewish law it is prohibited to enter the area where the Temple once stood in a state of impurity. It is for this reason that traditionally Jews pray at the Western Wall, just outside the area where the Temple stood.

The rules for the red heifer are very precise. The cow must be completely red (actually more like a ruddy brown color), (1) meaning that it does not have two hairs of any other color near each other or three non-red hairs anywhere on its body.(2) The cow must be at least three years old, have no physical blemishes, and must never have been used for labor. (3) Such conditions made the appearance of a proper red heifer very rare and in fact according to rabbinic tradition, throughout Jewish history only seven or nine were ever prepared. The most recent was prepared by the High Priest Ishmael ben Phiabi in the first century CE, shortly before the destruction of the Second Temple at the hands of the Romans.(4)

The red heifer ritual is generally considered the most mysterious rite in the Bible, an ordinance whose exact reasons are unfathomable. Rabbinic tradition teaches that even King Solomon, wisest of all men, was baffled by this ritual.(5) The early rabbis taught that this ritual was in some way atonement for the sin of the Golden Calf. (6) Some modern Bible scholars suggest that the idea of using a red cow for a purification offering has to do with the association of the color red with blood, thus symbolically increasing the amount of blood in the ashes. This would also explain the crimson yarn and reddish cedar wood that was burned along with the cow. (7) The main reason that the red heifer is discussed in contemporary news reports every so often is based on a tradition recorded by the medieval Jewish sage Maimonides. He writes that the tenth red heifer will be prepared by the Messiah.(8)

The ashes of the red heifer will then be used to purify people as part of the process of restoring the Temple service. It is because of this that the red heifer has taken on eschatological significance, its appearance understood as heralding Messianic times. It is for this reason that some people today search for the red heifer, even working to breed them.(9)

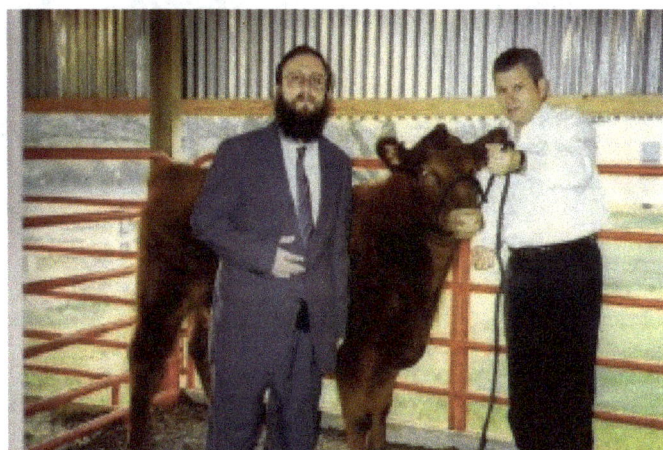

Fig 1

The Red

Heifer

The most recent attempts at finding a red heifer appropriate for ritual use include Rabbi Mordechai Shmaryahu of Kfar Chassidim in Israel, who in 1997 owned a newborn heifer named Melody that was originally declared a potentially kosher red heifer, but later some white hairs were found on her tail. When this red calf was first discovered, many were concerned that this would trigger some kind of activity to build the Temple in Jerusalem and upset the delicate balance among the various religious groups in Israel. Rabbi Shmaryahu himself demurred, saying, "The whole thing has been blown way out of proportion." (10)

The Temple Institute in Jerusalem, founded in 1987, is dedicated to rebuilding the Temple in Jerusalem. Most of their efforts centre on producing educational materials, raising public awareness about the Temple and constructing vessels and priestly garments for use in the future Temple. They run a small museum in the Old City of Jerusalem which is very popular with Christian tourists. (11)Over the years, rabbis of the Temple Institute have tried various ways of producing a kosher red heifer. One plan was to import frozen fetuses of a red heifer from a Scandinavian country and implant them in the uterus of an Israeli cow (12). Since the early 1990s Rabbis from the institute have been working with Clyde Lott, a Mississippi cattle rancher and Pentecostal preacher, to breed a red heifer based on Lott's Red Angus cattle. Like many Evangelical Christians, Lott hopes that the red heifer will herald the beginning of the End-Times. This activity has once again raised some concerns that it may be part of extremist efforts to destroy the mosque

Fig 2 The Third Temple as outlined in
Ezekiel chs. 40-47 Wikipedia Creative
Commons. Incorrectly labelled Second
Temple in Wikipedia.

currently on the Temple Mount as part of an effort to rebuild the Temple. The
rabbis from the Temple Institute caution that they have no intention to
immediately begin purifying people in preparation for the building of the Temple
in Jerusalem.(13)

Periodically, reports appear in the news about a red heifer being born. Already,
calves born through Lott's efforts have been declared candidates for a kosher red
heifer.(14) One was born in 2002 in Israel, but not much has been reported about
it since. (15) Because of the requirement that the cow must be at least three years
old for ritual use, in many potential red heifers white hairs develop later,
disqualifying the calf.

Rabbis from the Temple Institute explain that nowadays a red heifer is not that rare. (16) Herds of red cows are found in Holland and Texas, and recently the genetic code for red heifers has been cracked. (17) At this point the main problem is not finding an unblemished red heifer, these may exist already, and if not they will shortly. The main issue now is that the ashes of a red heifer can only be prepared by a priest who is himself ritually pure. However, the only way to be purified is through the ashes of the red heifer, resulting in a Catch-22. In Temple times, some ashes of previous red heifer were set aside and used to purify the priest who would prepare the ashes of the new red heifer so there was a solution to this problem. Until a mechanism is found to purify a priest who will prepare the ashes, the red heifer has no practical ritual use. In any event, as Rabbi Richman from the Temple Institute says, " Many things have to happen before the Temple can be built, the main one being that we must have unity in Israel before it can happen, which we don't have now. The Temple is supposed to be for everyone, not just Jews. It's really universal in scope." (18)

REFERENCES

1. Jacob Milgrom,ed., *The JPS Torah Commentary - Numbers* (Philadelphia: The Jewish Publication Society,1990), p.158.

2. Mishnah Parah 2:5.

3. Mishna Parah 1:1,2:2,2:3.

4. Mishna Parah 3:5

5. Eccles.Rabbah 7:23

6. J.H.Hertz, *Pentateuch and Haftorahs* (London: Soncino Press, 1965)p.652.

7. Milgrom, p.440.

8. Maimonides, Laws of the Red Heifer, end of chapter 3.

9. See my last article on the red heifer, " Watch Out for Red Heifer Madness," INFO Journal 78, Autumn 1997,p.23

10. "Holy Cow! Will Red Heifer Save World,Cattle Industry?" Jerusalem Post, July 11,1997," Letter From Jerusalem: Forcing the End" New Yorker,July 20,1998.

11. See http://www.templeinstitute.org/about.htm

12. "Speedily Rebuild,in Our Days." Ha`aretz, July 8,2008.

13. " A Very Holy Cow." Jerusalem Post, May 26,1997.

14. " The End of the World is Near to Their Hearts." Los Angeles Times, June 27, 2006.

15. "Red Heifer Days." National Review, April 11, 2002.

16. http://www.temple.org.il/show.asp?Id=38932

17. " Speedily Rebuild,in Our Days." Ha`aretz, July 8,2008.

18. " A Very Holy Cow." Jerusalem Post,May 26,1997.

The image on p 17 of the red heifer is from The Temple Institute web site.

The Pine Marten in Derbyshire 1996-2011

Richard Muirhead

According to `Evidence of Pine Martens in England and Wales 1996-2007:Analysis of Reported Sightings and Foundations for the Future` J Birks and J Messenger (2010) : " Of the period 1977-1982 *they* note: [Strachan et al-Ed.] "the number of records for the population of South Yorkshire/Derbyshire dwindled rapidly......,with little evidence of presence from West Yorkshire. There may have been some emigration southwards into South Yorkshire or the the High Peak area of Derbyshire. " (1) Subsequently (1983-1988), they suggest of this population: the paucity of records indicate that it may now consist of very few animals. (2)

VC 57 [Vice County 57-Derby-Ed] records are temporally concentrated in the 1990s, and geographically clustered in the centre of this VC in the wooded valleys between Bakewell and Belper (where SK35 is a hotspot hectad.) In 1996 a pine marten was seen 5m up a larch tree in Hall Dale Woods (SK2863), and another among rocks in Lea Wood near Holloway (SK3156); In 1997 (a year that produced six records from this VC) a pine marten was seen to run up a tree near Wirksworth (SK3254), one was watched for several minutes eating berries in a hawthorn bush at Crich (SK3454) and one was seen on top of a wall and then running up a tree at Wheatcroft (SK3557)And in 1998 a pine marten was described as `shooting up a tree in the Goyt Valley (SK0173); in 2002 one was seen 7m up a tree at Chellaston (SK32); In 2003 one was seen feeding on nuts up in trees and bushes Holloway (SK3156);and in 2005 a pine marten was seen moving from bough to bough > 20m up a tree near Eckington. (SK4179).Just before the publication of this report, a 2001 record of a pine marten killed accidentally during predator control in SK37.......came to light via Steve Docker of the Derbyshire Mammal Group. Photographs of the preserved specimen confirmed its identity, and we await a tissue sample to establish its haplotype.(3)

The Derbyshire Times of February 6th 1997 wrote: "Derbyshire naturalists are jubilant after the welcome return of one of Britain's rare elusive mammals- the pine marten. A sighting between Matlock and Belper, at a location which is being kept secret, could be the first in the county for more than ten years. Nick Moyes, who runs the Derbyshire Biological Records Centre at Derby's City Museum,said: "The animal was clearly observed for more than 15 minutes by naturalists who saw it feeding on berries high up in the branches of an overgrown hawthorn hedgerow." (4)

"The abundance of reports from the well-populated Derwent Valley over the period 1996-1999 is remarkable (with only four subsequent records).Some of these involved daytime sightings of animals close to human habitation, leading to suggestions that a release or escape from captivity may have occurred. The very heavy visitor pressure in the Peak District National Park (10 million visitors per year;ANPA,2009) would tend to ensure a reasonable recording effort within its bounds, , so the paucity of records from the new millennium may reflect a decline in in abundance"(5).

There were 25 records in the 1996-2007 period.

This VC produced two sightings of > 1 pine marten: in September 1996 two animals were seen together by a farmer herding cows near Riber near Matlock (SK3158); and in September 1996 three young animals were seen together on a road at night near Bradbourne. (SK2152) (6)

In April 2011 I sent a letter to the Buxton Advertiser on a purely speculative basis asking if anyone had observed pine martens in Derbyshire.I received two interesting replies. Mr Roger Leaning saw a dead one in the Summer of 2010 between the roundabout near the village of Baslow and the Robin Hood pub. This is a wooded area near Chatsworth House estate, with moorland. I contacted Chatsworth House in May 2011 to ask if the gamekeeper there had any reports of pine martens in their grounds but received the following reply:" Thank you for your email and I am sorry that you did not receive a reply to your earlier email. I am afraid that we do not have any records of sightings of pine martens in the Chatsworth area" (7)

One of the sightings was by a Mr Pete Yeomans:

Fig 1 Matlock area by Google Earth

returning to Parsley Hay and enjoying a four mile walk, me and the dogs. On crossing over the bridge that covers the road from the A515 Ashbourne road to Hartington (time about 8am), I spotted what at first I thought was a cat standing in the middle of the trail (roughly 4 yards from the bridge wall on the right hand side in the attached photo). It was staring at my two border collies and they likewise, staying long enough for me to get a good sighting of its markings. Not really 100% sure at the time what I had seen I checked on getting home and there it was on the good old PC a pine marten. Quite a sighting I was told later, an animal that is alleged to be extinct in this country. The one sighted disappeared quickly down the banking into the rough (on the right hand side of the photo), and I never saw it on my return from the signal box or ever again since, but I now keep an eye open when I am down any part of the White Peak Trails as you never know! Pete Yeomans (8)   See Fig 2 on page 24 for location of this sighting.

Fig 2 Location of the March 2008 sighting.

Mr Yeomans also sighted a pine marten in February 2011. Here is his account sent to me in the form of another e-mail:" Sighting of the elusive pine marten comes three years later in February 2011. Again down the old Victorian Tissington Trail, this time my walk was Hurdlow to Friden and return and on approaching a copse situated on the left side of the trail on a gradual left hand bend , between Cotesfield Farm and Parsley Hay I spotted what I now know was a Pine Marten. The photo is taken looking towards Cotesfield and the Pine Marten was on the right of the trail and on hearing or seeing me ran off to the right and over the wall and into what looks like a gully and thin copse never to be seen again. Lucky sightings or what from an animal rarer than a U.F.O. ? "(9) Pete Yeomans.

Fig 3 Location of the February 2011 sighting.

The Derbyshire Mammal Group News Spring 2011 issue contained this report:

Out with the old and in with the new? Genetic evidence of differing origins and fates of pine marten populations across the British Isles.

The Mammal Society Easter Conference April 2011

Neil R.Jordan, John Messenger,Peter Turner,Johnny Birks,Elizabeth Crosse,Catherine O'Reilly.

Question: Are relic pine marten populations still present in England and Wales?

"We investigated the origins and persistence of European pine marten (Martes martes) populations in the British Isles using mitochondrial DNA. Haplotypes of contemporary and historical marten populations from the same areas (Ireland and southern Britain respectively) differed. While Irish and southern British stock appear to have a common origin, the recent history of these populations differs. Genetic results from Ireland suggest that contemporary Irish pine martens are descended from a relict population which passed through an early 1900s bottleneck, while in southern Britain current data suggests a significant change in the population's genetic composition. In England and Wales, the apparently sole historical haplotype (I) has been replaced (since 1950) by a contemporary population consisting predominantly of individuals of haplotype A (currently also found in Scotland). This, and the occasional occurrence of haplotypes origination from continental Europe and others suggesting introgression with *M. Americana*, suggest that the relict populations of England and Wales have been replaced or at least infiltrated by occasional released, escaped and/or translocated animals. "(10)

Pine martens are alleged to have held out in the Derwent Valley until the 1970s. (11) .According to the Wild About Britain Forum in a posting by Ladywell on April 25th 2011 " For years there have been rumours of their [pine martens] existence in the Ladybower and Derwent woods but as far as I know, still nothing concrete as yet. Personally,I think it unlikely, but would love to be proved wrong.Someone in the Peak District once claimed they saw a giant slug as well, as big as a car and it might just be the same person who saw a pine marten.(12) Another contributor on April 27th reported a sighting above Stalybridge: " My brother was 90% sure he saw one in Snake woodlands about 2 years ago reported it to Vincent Trust."(13)

I have notes from the Victoria County History of Derbyshire vol 1(1905) on the occurrence of the pine marten in Derbyshire in the 19th century which I can pass on to any reader for free if you interested. Just give me your e-mail address.

REFERENCES

1. J.Birks and J.Messenger citing Strachan et al in Evidences of Pine Martens in England and Wales 1996-2007:Analysis of Reported Sightings and Foundations for the Future p. 81

2. Ibid p.81

3.Ibid p. 81

4. The Derbyshire Times February 6th 1997.

5. J.Birks and J.Messenger op cit p.82

6. J.Birks and J.Messenger op cit p. 81

7 E-mail from Chtasworth House to R.Muirhead.May 2011

8 E-mail from Pete Yeomans to R. Muirhead April 29th 2011

9. E-mail from Pete Yeomans to R. Muirhead April 29th 2011

10 Anon Derbyshire Mammal Group News Spring 2011 p 1

11 P.Hobson The lost animals of Derbyshire. Derbyshire Life and Countryside vol 69 no. 12 December 2004 p.75

12 Posting by Ladywell on Wild about Britain Forum April 25th 2011

13. Posting by JB3902 on Wild About Britain Forum April 27th 2011

The Weird Weird World of Worm Anomalies

Richard Muirhead

Job ch 25 vs 5-6: "Behold, even the moon has no brightness (compared to God`s glory) and the stars are not pure in His sight. How much less a man,who is a maggot!And a son of man,who is a worm!"

This essay does not pretend to be an exhaustive coverage of worm anomalies,just some highlights I found interesting:For a creature as humble as a worm, there is a fairly significant quantity of unusual data. One of my earliest pieces of reading material was about Lowly Worm and his chums in the Richard Scarry books I read as a child in Hong Kong. My Dad found me playing with a " family of worms" once,this was also in Hong Kong. So I decided to gather together the following stories which are arranged roughly in date order. This first selection relates to worm distribution in the Sahara and beyond from `Sahara: The Life of the Great Desert`. (2004)

"Desirable as they are, earthworms are also vulnerable creatures. They are extremely sensitive to the loss of body water, and will die within hours if exposed to the air,particularly aquatic species. Yet earthworms exist in all the damp places of the Sahara, and aquatic worms in oases a very long way from places where rain commonly falls or rivers commonly run - in Tin Téhoun, for example.Tin Téhoun is a small hamlet, not much more than a few ramshackle houses and the well, shabby and down-at-the-heels. Even the palms look dispririted, poor cousins to the robust groves of the northern oases. The nearest occasional water, the Niger River floodplain, is at least fifty miles away to the south. To the north, sixty or seventy miles away, is another hamlet not much different. Maybe two hundred miles to the north-east is the drainage basin around the Timétrine Mountain, a flattened adjunct of the Adrar des Iforhas, but the drainage basin hasn`t actually drained anything for millenia, and is as arid as the surrounding desert. Yet here are earthworms, as slimy and glossy as ever.

How did they get here? Where from? Why here? How long ago? Did they migrate here, in different times, or are they retreating?Were they always here, and are they now cut off from their "cousins" elsewhere? And if they did arrive in the deep desert in earlier epochs, when? In any case, their "migration routes" shed some light on geological events and on the history of human interaction with the Sahara, as an Egyptian team thought when they began an esoteric and unlikely study of deep-desert earthworms.

The Egyptian Study,by S.I. Ghabbour of Cairo University's Institute of African Research and Studies,had become interested in genus,and then in species. Earthworms may look alike,but different parts of the Sahara are inhabited by different species,with some curious connections.. The worms of the north-west-Western Sahara, Morocco,parts of Algeria, even the offshore Canary islands-are identical to those in Spain and Portugal. Species found in eastern Algeria,the Tripoli area of Libya,and as far east as Siwa in Egypt are similar to those found in Sardinia and Sicily.

The worms of eastern Nile region can also be found in the Levant and as far north as Romania.The earthworms of the southern Sahara live along four distant tracks. The first follows the Blue Nile north from the Ethiopian Plateau;the second starts in Kenya and ends at the Siwa oasis in northern Egypt;the third starts near Africa's Great Lakes, especially around Lake Victoria,passes through Sudan,and finishes in Tunisia;the fourth,most curious of all,is a long and sinuous trail that starts in West Africa,in Liberia and follows the southern Sahara all the way to the Nile and thence north to Lake Dahshour,near Cairo, a distance of more than three thousand miles.

The migration patterns, if that's what they are,of the Saharan earthworms in some ways confuse as much as enlighten.But their widespread dispersal in the desert is clear and unambiguous evidence that the Sahara was once a great deal more humid than it is now,at some or many times in the distant past. The earthworms aren't by any means the only evidence,but they are conclusive. The earthworms didn't hitch a ride on a camel;they came on their own,through soil that was moist and nutritious,and were trapped in the few places where moisture remains."(1)

Fig 1 Page 30. A very odd worm from The Country-side Monthly September 1910 vol 1 no. 4 page 166.

Worm With Two Tails.— I thought the sketch of this extraordinary "earthworm freak" which recently came into my possession, might be interesting to readers of THE COUNTRY SIDE. It was dug up by a relative, who wanted some earth worms as bait for fishing, and had dug up some hundreds before for the same purpose without ever seeing one like this. The curious way in which the worm spreads out may be seen from the enclosed sketch and one fact worthy of mention is that it shows no sign of any injury and is still being kept alive. The sketch is natural size.—A. DOROTHY DAFFON, Cirencester.

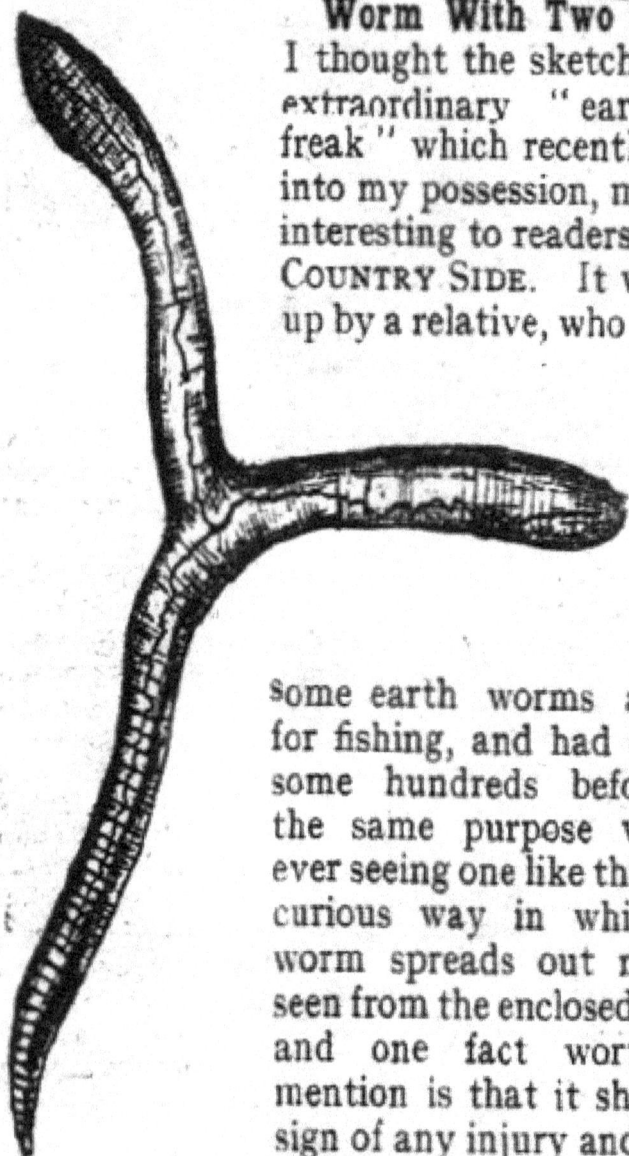

[Similar freaks are not unknown, but are of very rare occurrence.—E. K. R.]

The Countryman magazine for Summer 1993 contained the following contribution from Don Chapman of Harlow,Essex:

THE NIGHT THE WORMS DANCED. It was 2am on a warm,summer night,very humid with the heaviest dew I can recall. There was a full moon in a clear sky. Drawn by the brilliant light outside, I got out of bed, opened the window and looked onto the lawn. It was covered by hundreds, thousands even of very large worms spaced out evenly all over its 20 square yards,each worm aligned towards the moon which was high in the southern sky. I went out and walked among them. As I approached, the nearer ones withdrew at the last moment in a smooth, unhurried fashion and I realised that although each lay easily nine inches along the ground,it had its tail end inside the hole it had appeared from, giving purchase on the sides and enabling rapid withdrawal. Those farthest away remained as they were, saturated in dew and moonlight as if basking. I got up again later in the night and looked out again to find the worms still aligned to the moon which had now moved round about 30 degrees. Like so many moondials, they pointed in unison at the moon. They have no sight, so how could they have known where was? This event has remained in my memory for 20 years, I have never seen it repeated nor heard of it elsewhere-*Don Chapman,Harlow,Essex*

Fig 2 Earthworm Wikipedia Creative Commons

[Earthworms do have light-sensitive cells. They recoil from strong light but are attracted to weak light, which perhaps explains apparent orientation to the moon. Over 150 years ago, Gilbert White described other aspects in *The Natural History of Selborne*: 'When earthworms lie out a-nights on the turf, though they extend their bodies a great way, they do not quite leave their holes, but keep the ends of their tails fixed therein, so that on least alarm they can retire with precipitation under the earth. Whatever food falls within their reach when thus extended , they seem to be content with, such as blades of grass, straws, fallen leaves, the ends of which they often draw into their holes....'- E.D.] (2)

I also have a worm story from Mars from Fortean Times 139 (October 2000) pp 24-25. This article is 'Mars Special: Water, Water, Everywhere' in which reports are made of worm-like structures on Mars. " Skipper's own interpretation of the 'worm-like tubes is quite extreme he concludes that they might be "organic constructs" that capture and retain water and may still be functioning. "(3) Skipper and fellow Mars watcher Jeffrey McCann now have their own website marsanomalyreserach.com to publish their discoveries.

Fig 3 Animals and Men 3 1994 showing very long worm. Reproduced with permission of Eastbourne Gazette and Jon Downes.

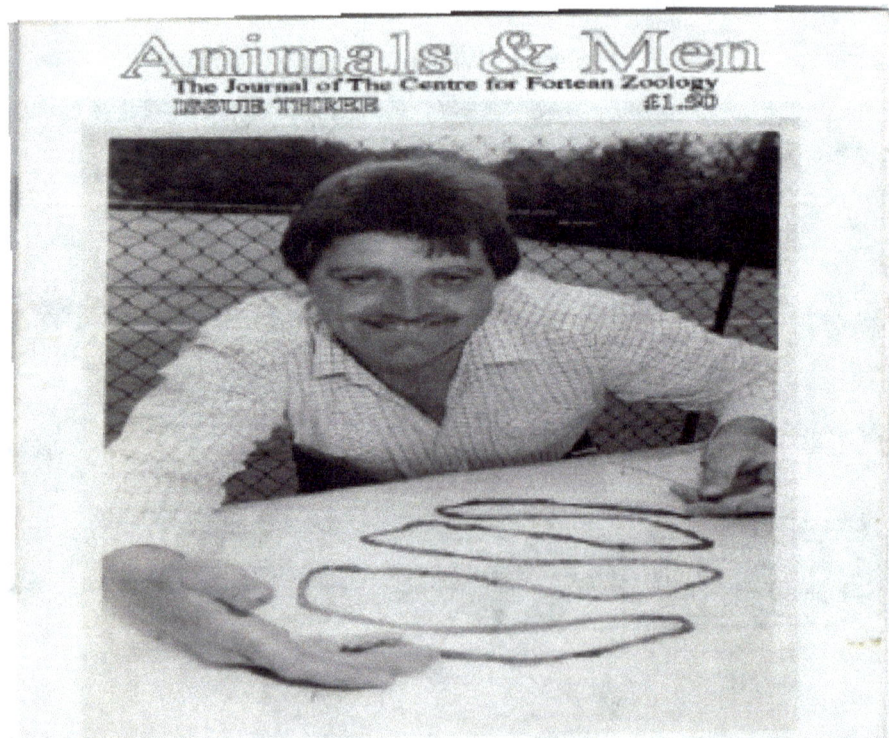

The account of the giant worm on the previous page in Animals and Men 3 said:
"Although the creature was of local interest as a `freak` it was not a record holder
(the British record Earthworm was 13 feet in length and the world record is held
by a South American 33 footer) after being photographed this 20th century
descendant of ` Oroborous the Midgard Serpent` was consigned ignominiously to
the rubbish skip! "(4)

Charles Fort obviously had something to say about worms, as follows:

"That, June 24,1911, at Eton,Bucks,England, the ground was found covered with
masses of jelly, the size of peas, after a heavy rainfall. We are not told of
nostoc,this time: it is said that the object contained numerous eggs of "some
species of Chironomus, from which larvae soon emerged."

"I incline,then,to think that the objects that fell at Bath were neither jellyfish nor
masses of frog spawn, but something of a larval kind- This is what happened in
Bath, England,23 years before. London *Times,* April 24, 1871: That,upon the
22nd of April, 1871, a storm of glutinous drops neither jellyfish nor masses of
frogspawn, but something of a [line missing here in original text. Ed.] railroad
station, at Bath. " Many soon developed into a wormlike chrysalis, about an inch
in length." The account of this occurrence in the *Zoologist* , 2-6-2686, is more like
the Eton-datum; of minute forms, said to have been infusoria; not forms about an
inch in length." (5)"........In Timbs Year Book , 1877-26, it is said that, in the
winter of 1876, at Christiana, Norway, worms were found crawling upon the
ground. The occurrence is considered a great mystery, because the worms could
not have come up from the ground, inasmuch as the ground was frozen at the
time, and because they were reported from other places,also, in
Norway..........Large number of worms found in a snowstorm, upon the surface of
snow about four inches thick, near Sangerfield,NY Nov 18, 1850 (*Scientific
American* ,6-96) . The writer thinks that the worms had been brought to the
surface of the ground by rain, which had fallen previously"(6)

New Orleans Daily Picayune, Feb. 14, 1892-enormous numbers of unknown
brown worms that had fallen from the sky, near Clifton,Indiana. *San Francisco
Chronicle,* Feb. 14, 1892-myriads of unknown scarlet worms-somewhere in
Massachusetts-not seen to fall from the sky,but found,covering several acres,after
a snowstorm. It is as if with intelligence, or with the equivalence of intelligence,
something has specialized upon transporting, or distributing ,immature and larval
forms of life. If the gods send worms, that would be kind if we were robins. (7)

On April 29th 2010 The Metro newspaper reported:

Unearthed: Worm not seen for last 20 years

"An earthworm that has acheieved near-mythic status in the U.S. Has been found alive. An adult and juvenile giant Palouse earthworm were found after scientists used a probe that sends electric shocks into the ground, forcing them to surface. It is the first time in 20 years that two live specimens have been found. The worm- the latest example of which measures 30cm (12in) - was first reported to the scientific world in 1897 but largely disappeared from the Palouse region of Washington and Idaho for nearly a century. Some experts blamed agricultural development. The latest pair were found by Idaho University researchers who dispelled two bizarre myths that surround the creatures: no, they did not spit and they did not smell of lilies. The two were found last month but were later identified as the elusive species. Three cocoons were also found, two of which have hatched."(8)

REFERENCES

1. M. De Villiers and S. Hirtle Sahara: The Life of the Great Desert (2004) pp 38-40

2. The Countryman Summer 1993 p.75

3. Fortean Times 139. October 2000 p.25

4. Animals and Men 3. 1994 p. 19

5. The Complete Books of Charles Fort. The Book of the Damned p. 48 (1974)

6. Ibid. The Book of the Damned p. 97

7. The Complete Books of Charles Fort Lo! p 548

8. The Metro April 29th 2010.

RANDOM

DEVO IMAGE!

Some Odd Fauna Records in Britain 1611-1922

Richard Muirhead

John Speed`s ` **Theatre of the Empire of Great Britaine**` was first published in 1611 so it seems appropriate to start with that year in this study. This was a kind of atlas with notes about each county. For the Whitby area of Yorkshire Speed states: " Places of memorable note are *Whitby,* where are found certaine stones fashioned like Serpents, foulded and wrapped round in a wreath, even the very pass times of Nature, who shee is wearied with serious workes,sometimes forgeth and shapeth things by way of sport and recreation: so that by the credulous they are thought to have beene Serpents, which a coate of crust of stones had now covered all over, and by the praiers of *S.Hilda* turned to stone: And also there are certaine fields here adjoining, where Geese flying over fall downe sodainlie to the ground, to the great admiration of all men: But such as are not given to superstitious credulitie, may attribute this unto a secret propriety of this ground, and a hidden dissent betwixt this Soile and these Geese, as the like is between Wolves and the Squilla rootes. At *Skengraue* (a little village) some seventie yeeres since, was caught a fish called a Sea-man, that for certaine daires together fedde on raw fishes, but espying his opportunitie escaped againe into his waterie Element....At *Huntly Nabo*, are stones found at the rootes of certaine rockes, of divers bignesse, so artificially shaped round by nature, in maner of a Globe, as if they had beene made by the Turners hand. In which (if you breake them) are found stony Serpents, enwrapped round like a wreath, but most of them headlesse"(1)

"In England the folklore of snakestones is centred mainly around Whitby in Yorkshire and Keynsham in Somerset. From near Whitby William Camden (1551-1623) in his Britannia of 1586 recorded stones which `if you break them you find within stony serpents, wreathed up in circles, but generally without heads`.Legend supposes that the fossils were once living serpents which were common in the area until the 7th Century A.D when Saxon abbess St Hilda (614-680) turned them into stone in order to clear a site for the building of her convent. The heads of the serpents were assumed to have been destroyed on their death. This legend has passed into English literature as illustrated by the following passage from a poem by Surtees:

"Then sole amid the serpent tribe,The holy Abbess stood,

With fervent faith and uplift hands

Grasping the holy rood.

The suppliant's prayer and powerful charm

Th'unnumbered reptiles own;

Each falling from the cliff, becomes

A headless coil of stone "(2).

A similar passage from Sir Walter Scott's *Marmion* describes the same tale:

"When Whitby's nuns exalting told,

Of thousand snakes each one

Was changed into a coil of stone,

When holy Hilda pray'd;

Themselves, within their holy ground,

Their stony folds had often found".(3)

The absence of heads in the Whitby snake stones is sometimes attributed to a further curse by St. Cuthbert, another saint from northern England. In order to perpetuate the legend and to effect sales of specimens, local collectors and dealers in fossils frequently 'restored' the snake stones by carving heads on them. Many of the Whitby fossils were preserved in jet, which when carved and polished could make beautiful ornaments; the Vikings imported jet from Whitby, and at least one carving of an animal resembling a snake stone is known from Norway, while in Elizabethan England snake stone brooches of jet were highly prized." (4) The front cover of FS 2 shows a snake stone- a fossil ammonite.

There is a story, possibly quite ancient of " a dry land grampus, living in a yew tree", whatever that may be, at a village near Highclere in Hampshire , in John Edgar Mann's **Hamphire Customs, Curiosities and Country Lore.** Now a grampus is a " a killer whale or other cetacean of the dolphin family." (5) so what one was doing in a yew tree in Hampshire I do not know, perhaps the original eye witness was on some sort of hallucinogenic drug?! Moreover , "the vicar, with bell, book and candle, managed to banish [it] to the Red Sea for a thousand years.....(6) So if you are reading this now and the 1000 years are up and you look out of the window at that yew tree, take care! Seriously- Richard Freeman has said: " Once Nile crocodiles ranged as far north as Turkey. There is even a river called the Crocodile River in Turkey. I don't know if they were as far north as the Red Sea though.I heard of a case in Africa where a crock bit a chunk out of a diver's hip. The guy got away and climbed a tree and the croc clambered 3 metres into a tree after him. That's the only account I can recall."(7)So it's unlikely but not impossible that this cryptid was a crocodile and there is no river in Highclere itself. Although 'Highclere' itself means ' bright river.' The nearest river is the Endbourne in between Highclere and Newbury. Perhaps on one occasion there was a flood that washed a cetacean into the branches of a yew tree? Perhaps it was a kind of giant slug? Ron Snipp of Highclere History Group wrote to me: " A couple of lines regarding the Highclere Grampus are below - of course we have heard of this story but are not sure when it will be free to return - those of us who visit the church,live in trepidation and hope that this act was at the old church that originally stood in the Castle grounds! "(8) Also, this information from the **Highclere Village History Group** web site of February 26th 2001: 'The Highclere Grampus - anything known? :

A web site specialising in dragons tells us that our village is home to 'The Highclere Grampus.'

"A grampus being a word describing a killer whale, which nestled in a yew tree. It was banished into the sea for 1000 years by a parish priest. "

HIGHCLERE Here another beastly legend is based. Near the church at Highclere there is a Yew Tree in which lived a Grampus, which is a beast something like a Dolphin, except that it lives on the land instead of the sea. This beastly creature has the reputation of "blowing like a whale", which may account for the expression " puffing like a Grampus".

The Grampus at Highclere was known to be rather timid and not very fierce, but it occasionally scared the villagers with its noisy breathing, chasing those who referred to it as "snoring"! (9)

" One of the few printed references to the Grampus was published in 1890, in Andrew Lang's " **Life,Letters,Diaries of Sir Stafford Northcote- First Earl of Iddesleigh.**(vol 1 p.220) Northcote, who lived between 1818 and 1887, was a British Conservative politician and lover of tales of the paranormal. He chronicled a short passage about the Grampus - which refers to as a " Grumpus"- in his diary:

" A few days after autumn he spent in Highclere. His shooting 'was excerable', but he was consoled with an evening of ghost stories. `Mrs - had the advantage of us in having seen a ghost.` He expected a visit from Grumpus, the Highclere bogy, who,it is true,had been laid in the red sea for 100 years, but his time there was now nearly expired."

While this brief account does not contain any new information on the Grampus, what it may help to illuminate is what period of time the beast actually lurked in the yew tree in the Highclere Churchyard.

Northcote indicated that this fiend's banishment lasted merely 100-years - which may be a typo - but most accounts agree that this beast was cast out for no less than millenia. Assuming that the 1000 year exile is accurate, that would place the Grampus's tenure in the yew tree in Hampshire County at somewhere in the in the order of the late 9th century A.D.(10)

Fig 1 Grampus or Orca

Wikipedia Creative Commons

"

On April 27th 2011 Marco Masseti (see Flying Snake 1) sent me the following information in response to queries about entries in John Fleming's History of British Animals (1828). One of my questions was about the " Cypress Cat". Another question was about the Beech Marten in Britain. (see p. 45)

FELIS.Cat. " The spotted variety, termed "Cypress Cat"[1],is noticed by Merret,[Christopher Merret,1614/15-1695, an English physician and scientist who compiled one of the first lists of the flora,fauna and minerals of Britain, the *Pinex Rerum Naturalium Brittanicarium* - see Wikipedia, Richard] who says (Pin.169.) "Enutritur in aedibus nobilium [2] ." I haven't been able to translate this on any Google translation tool, nor can I find Cypress Cat on a Google search.

---

[1] I cannot understand why they named this cat as "Cypress Cat"The term may underline the fact that it was an exotic element to the biogeography of the U.K. Like the cypress that is extraneous to the natural vegetation of Great Britain and Ireland. In any case , one might refer this spotted cat to many wild large-medium sized felids of the Old and the New Worlds, such as the cheetah, *Acinonyx jubatus* (Schreber,1775) ,the Iberian lynx, *Lynx pardinus* (Temmink,1827), the African serval, *Leptailurus serval* (Schreber,1776), several Asian representatives of the genus *Prionailurus Severtzov*, 1858, and/or of the American genus *Leopardulus* Gray , 1842,etc(see also Masseti, 2009)

Masseti M., 2009- Pictorial evidence from medieval Italy of cheetahs and caracels, and their use in hunting. *Archives of natural history*, 36(1) : 37-47

[2] *It feeds* (or,not literally: *"it is bred"*) *in the houses of nobles*. This means that it was regarded as a very precious (and,perhaps exotic) animal!

The early 18th Century has a number of interesting records, this from **The Lancashire Journal** February 5th 1739: " There was taken of the Hackney River at Mrs Smith's Preston Ferry, near Clapton, a monstrous Creature of a Fish, which has four Eyes,its Head like a Jack, two Arms like a Child,paw'd like a Bear, Claws like an Eagle,and a Tail like an Eel, a Crown on his Nose, and is six Foot in length "(11)

The following item is from **The Lancashire Journal** of May 29th 1739 in Kent: "We hear from Dover, that some Days ago Mr Kennet a Miller of that Place, catched a Trout weighing three Pounds and three Quarters; but what is very remarkable, it being affirmed for Truth, that he found three Water Rats in its Belly, and a fourth almost digested."(12)

Fig 2 Brown Trout Wikipedia Creative
Commons

The Lancashire Journal for January 21st 1740 reported: " Last Thursday a pair of very large black Eagles settled on an Island called the Binness, belonging to the Estate of Colonel Smith. They seemed to be about Four Feet high, as they stood on the Ground. The appearence of these Birds which are very rarely seen in this Country, happening to be observ'd the next Day after the Publick Fast for imploring Success on the present War, occasions various Speculations, and is look'd upon as a happy Omen of the Reunion of the Imperial Arms of Germany and Great Britain, in order once more to check the growing Power, and humble the exorbitant Pride of France and Spain"(13)

Andrew Judd a contemporary Macclesfield ornithologist examined these birds and commented as follows: " I've found no record of Eurasian Black Vultures in the UK.The official British bird list held by the British Ornithologists' Union (B.O.U.) only goes back to 1800. They are in the process of producing a list of earlier species seen in Britain.

The British List is at http://www.bou.org.uk/thebritishlist/British-List-2010.pdf

However having looked up the Eurasian Black Vulture (note " Eurasian"- There is also a species in America known as a Black Vulture) in my bird books and also on Wikipedia there are several things that make this the most likely suspect.

1. They are dark with the adults having pale heads. However the juvenile is totally black.

2. Though they have not been recorded since 1800 their range traditionally covered Portugal, Spain and southern France. Birds from considerably further afield have been recorded in Britain. It is not impossible for a storm to have caused a couple to have landed in Britain or possibly for a juvenile or two to have wandered from their usual grounds. They are not however migratory and most rarities in the UK are from birds that have been blown off course during migration.

3. They are considered to be the largest bird of prey in the world, a good point considering the size of the reported sightings. The length of the bird, considered to be from the tip of it's tail to the tip of its beak when laid out flat is up to 115cm (45.3 inches).While they would not stand so high it's close to the four feet the observer estimated making allowances for some error.

So my conclusion is that the sighting was of two juvenile Cinereous Vulture or Eurasian Black Vulture. Latin name is Aegypius monachus. "(14)

Fig 3

Eurasian Black
Vulture.
Wikipedia
Commons

In the **Manchester Magazine** June 24th 1740: " They write from the Country, that an *Insect*, which first made its Appearance in *Norfok*, of the Caterpillar-kind, with a formidable Pair of Forceps,Harpy Talons, and a blue List a-cross his Belly, has fastened on the Herbage, which it destroys to the very Root: So that in some Parts *Forrage* is like to become as scarce, as if it had been all engrossed for the Use of the *Standing Army*.(15)

Fig 4 on page 44 shows an impression of what this insect might have looked like. Fortunately agricultural records are available for 1740 which indicate the kind of conditions then prevailing, as follows: " A year of drought, the beginning of a spell of four exceptionally dry years..........It was still cold in July,and the rainfall figures were redeemed largely by a very heavy thunderstorm in late July. Harvest was late and poor, and much fruit failed to ripen. Gilbert White records that fieldfares remained in England till June......A hurricane occurred in London on November 1st. Jethro Tull died.(16)

On August 14th 1778 **The Belfast News-Letter** had the following intriguing report, one of the earliest accounts of a British crocodile (if indeed that was what it was): " A few days ago was killed at Coatham,near Kirkleatham,in Cleveland, a very extraordinary monster that resembled a Crocodile. It was seven yards long and was thought by a numerous company, who assembled to see it, to be the most surprising creature ever seen on the coast of England." (17)

Enquiries at a museum in Kirkleatham in 2010 yielded no information about this "crocodile."

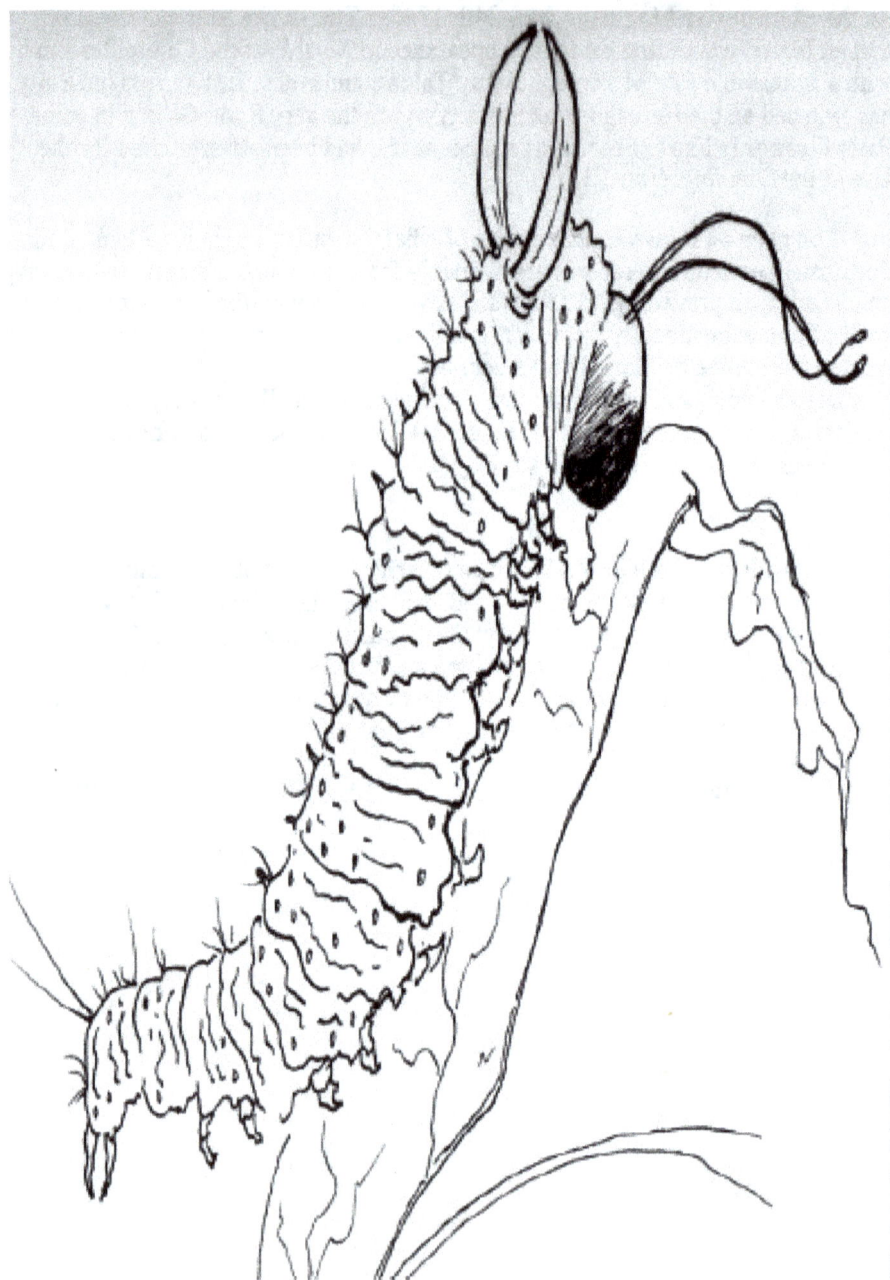

Fig 4 Unknown insect in
Norfolk, 1740, as portrayed by
Mike Hardcastle.

Marco Masseti sent me the following additional information in response to my queries about entries in John Fleming's **History of British Animals.** (1828)

MARTES. " M.fagorum. Common Martin" (i.e The Beech Marten [1]) - throat and breast white.........In woods and rocks in the south of England.[possibly confirming Jon Downes's view in **The Smaller Mystery Carnivores of the West Country** (2006) of the presence of the Beech Marten there, but see footnote 1 below.-R]. The length of the body is about 18 inches, the tail 10. The general colour of the fur is dark brown, the head having a reddish tinge- It is a great destroyer of poultry and game. Easily tamed. Lodges frequently in hollows of trees, and brings forth from four to six young." (18)

Note 1 below is completely challenged by Jon Downes research in his book **The Smaller Mystery Carnivores of the West Country** which shows that the Beech Marten and Pine Marten have survived in the S.W of England until relatively recently and indeed may still survive.

Fig 5
Enormous
horns, John
O' Groats
Journal
1840

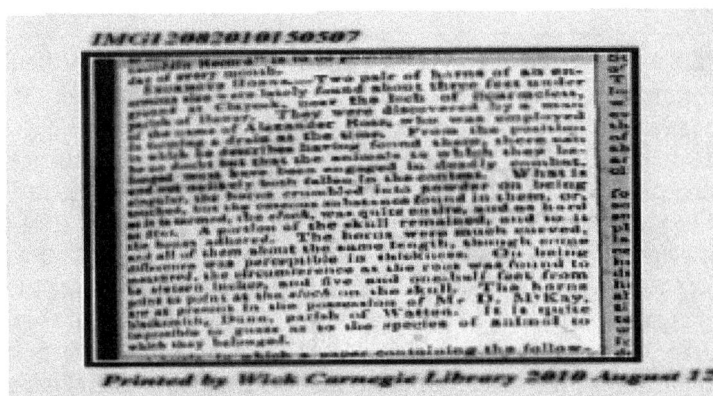

The Countryman for Summer 1975 includes the following :"I was reminded ...of an entry in the Churchwardens' Accounts for the village of Lythe, near Whitby in North Yorkshire, recording that, in 1846, 8s was paid for 'One jackall head'. For this I have not been able to find an explanation but the high price suggests the animal was an unusual one." (19) For a dog-fox or dog-jackal hybrid in 1907 see p.47.

---

[1] There are no beech martens or stone martens, Martes foina (Erxleben, 1777), in the UK, but only pine marten,Martes martes L., 1758.According in fact to Yalden (1999), the evidence from Europe is that the beech marten was a late immigrant to the West: It was possibly a follower of human (Neolithic) groups. Kurtén (1968) observes that it was common in the Near East in the Late Glacial, but was probably absent from western Europe.

The Nottinghamshire Guardian contained the following item about a racoon in Britain, in its May 7th 1857 edition, an animal that has appeared up to the present day: " A NOVEL FISHER - On Tuesday last, Mr Neilson,fisherman,Auchencairn,on visiting his salmon.nets, was suprised to find an animal, to him unknown, busily engaged in devouring fish. On the east coast seals are great pests to the fishermen, but on the shores of the Solway Firth they are unknown;pellochs are common enough, and very destructive to salmon,but,as far as we know, they never come within the dangerous precincts of stake-nets. After a severe struggle, Mr Nelson,aided by his dog,succeeded in destroying the interloper, and sent its carcass off to our ingenious townsman, Mr Hastings, to be stuffed. Mr Hastings at once recognised the animal as a racoon, and undoubtedly it is a very fine specimen of the common racoon (ursus lbtor of Linnæus), a native of North America."- *Dumfries Courier.* (20)

In 1887 the following story relating to giant pike in Market Drayton, Shropshire, appeared in The Nottinghamshire Guardian`s Natural History Column for July 16th 1887:

The following is taken from the **Live Stock Journal**, and will no doubt be interesting to our piscatorial readers:- " As event occurred recently at Market Drayton of so singular a character that, were it not authenticated beyond the shadow of doubt, I should not venture to narrate it. Some months ago the butler at Combermere mysteriously disappeared, and it was generally believed had been drowned in the mere. To solve if possible the mystery of his fate, the services of a diver were secured, who has been busily secured, who has been busily engaged in exploring the depths of the mere. This is a stretch of water between 150 and 200 acres in extent, varying in depth from six to forty feet. It has long been a noted place for pike, and the diver affirms it is positively bristling with them, and that so little did his appearance in their domain affect them, that several have glided so close past him as to brush his legs with their tails in passing. One day,however, when in about eight of water, a monster darted from the shadow of an old tree root, and struck him full in the chest, knocking him on his back, causing him so much pain that he became sick and faint, and had to return to the surface with all speed. He said he believed the fish had taken his arm in its mouth, but it is thought that the fish struck it with its tail, and thus injured it, as no marks of teeth were visible. The diver said, in a long experience both in sea and river, never had he been so assailed before, and he took care to provide further protection for his hands and arms before proceeding with his task. " (21)

The Country-side reported on December 9th 1905 page 55 an odd coloured frog in N.Wales. The frog was described as having a back of a bright red colour, spotted with orange, and a pure white breast seen swimming in a brook in N.Wales, it was an interesting instance of colour variation of the common frog. On the same page a bright yellow frog in Hull was mentioned. The Country-side September 1st 1906 p. 230 Grass snake 2ft long killed in Ballymena, Co. Antrim. The same magazine on September 28th 1907 (p.294) contained the following: Dog-Fox Hybrid?- " This curious animal, supposed to be a cross between a dog and a fox,was killed some time ago wild, in a wood in Warwickshire. In colour and shape it resembled the fox very much,especially the hind quarters, as will be

in the photo, the tail is thick and marked at the tip with white the same as a fox.When killed, it had very much the same scent;in size,not quite so large as a fox.- ARTHUR QUATREMAIN (22)

Fig 6 A dog-fox hybrid Warwickshire 1907

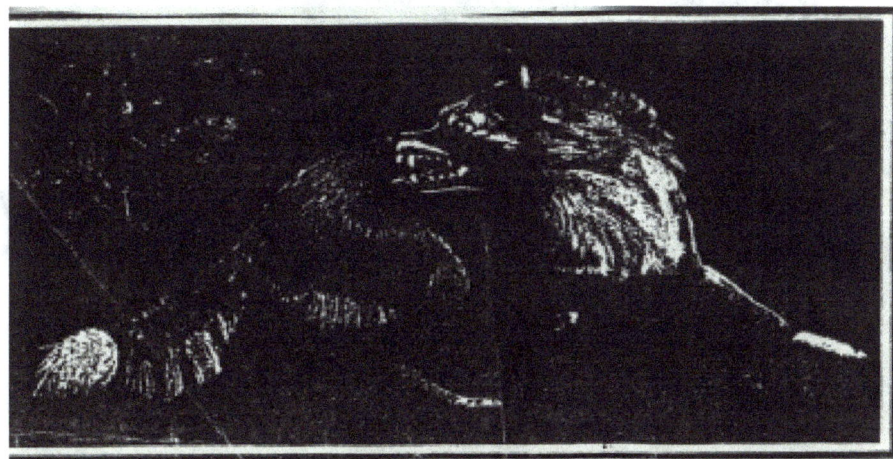

A Curious Animal.
This creature, supposed to be a cross between dog and fox, was killed in Warwickshire.

[A. Quatremain

[Terry Hooper wrote to me on August 12th 1997 saying: " Karl Shuker tells me its more likely a dog-jackal would mate (but,then,my accounts state the jackal owner in one case bred a Jackal vixen with fox?)..........It's curiously fox/jackal headed (more fox?) but collie dog bodied."(23)]

On July 19th 1912 the following story and photograph appeared in the Northampton Mercury featuring a "Tasmanian cat." The animal is of course the Madagascan ring-tailed lemur, so what was one doing near Weedon, Northamptonshire in 1912 and why was it thought to have come from Tasmanaia?

Fig 7

"Tasmanian

Cat"

The photograph is of a Tasmanian cat found on the line near Weedon, and given by the railway officials in charge of Mr. B. Southgate, of the Horseshoe Inn, Weedon, until the owner can be found. Mr. Southgate is making use of the opportunity by collecting for the Northampton Hospital.

In 1912 **The Country-side** reported a Green lizard in Lancaster p. 410

Fig 8

Rabbit with

tusks

The headline in this February 1922 edition of **The Surrey Times** reads `Rabbit With Tusks`.Obviously a sad deformity.The text reads: "The above is a photo of a rabbit killed at Hascombe last week. It will be noted that the lower jaw protruded in front of the other, with the result that two of the teeth grew like tusks and penetrated the upper jaw. The mouth was practically sealed , and when killed the animal was but a skeleton". (24)

REFERENCES

1. J.Speed Theatre of the Empire of Great Britain 1611 Folios 75-76.

2. M.G.Bassett `Formed Stones,` Folklore and Fossils 1982 pp. 4-5.

3. M.G. Bassett ibid p. 5.

4. M.G. Bassett ibid p.5.

5. Concise Oxford English Dictionary (11th ed) 2008 p.618.

6. J.E. Mann Hampshire Customs,Curiosities and Country Lore (1994) p. 13.

7. E-mail from Richard Freeman to R. Muirhead June 13th 2011.

8. E-mail from Ron Snipp to R. Muirhead July 7th 2011.

9. www.southernlife.org.uk/folklor3.htm.

10. http://americanmonsters.com/site/2010/10/grampus-england/.

11 Lancashire Journal February 5th 1739.

12 Lancashire Journal May 29th 1739.

13. Lancashire Journal January 21st 1740.

14 E-mail from A. Judd to R.Muirhead June 29th 2011.

15. Manchester Magazine June 24th 1740.

16 J.M. Stratton Agricultural Records p. 75.

17 The Belfast News-Letter August 14th 1778.

18 E-mail from M. Masseti to R.Muirhead April 27th 2010.

19. The Countryman Summer 1975 p.186.

20 The Nottinghamshire Guardian May 7th 1857.

21.. The Nottinghamshire Guardian July 16th 1887.

22. The Country-side   September 28th 1907.

23.   Letter from T. Hooper to R.Muirhead August 12th 1997.

24   Surrey Times February 1922.

A UNICORN RABBIT FROM COUNTY DURHAM

Karl Shuker

As someone passionately interested in unicorns, I have documented many different types over the years, ranging from the familiar equine version to rather more exotic counterparts - including a lethal carnivorous desert dweller with a musical flute-like horn, an ostensibly semi-aquatic form with webbed feet, an extremely bellicose bovine or even rhinocerine equivalent from Persia that could be soothed only by the calming cooing of a turtle dove,and a small hare-like but extremely malign entity from an unnamed tropical island. However, there is one particular example, which I investigated a fair few years ago but have never previously documented, that I find especially intriguing - for the simple reason that whereas most unicorns of whatever type they may be are fictitious, this one was real.

On 29 September 1982, writer Paul Screeton at the *Hartlepool Mail* published a report (subsequently picked up by other media sources, and also reproduced in his own magazine, *The Shaman* - see photo) documenting a most extraordinary pet rabbit that its owner, 9 - year- old Kathy Lister of Trimdon Grange in County Durham, England, had very aptly named Unicorn. Due to a genetic fluke, Unicorn had been born with just a single ear. Yet whereas there are numerous reports on file of individual mammals of many different species in which one or other ear is missing, Unicorn's condition was rather more special. For unlike typical one-eared individuals, her single ear was not laterally positioned, but arose instead from the centre of her head, standing upright like a long furry horn!

Intrigued by this highly unusual condition (even today, I have never encountered any additional `median-ear` instances), I decided to pursue the case personally. So after first discussing it with Paul Screeton, in July 1988 I contacted Kathy (then aged 15) and her father James, requesting further details, and am most grateful for the following information they very kindly sent to me.

Born in Spring 1981,Unicorn was a Flemish Giant doe bred on James's farm, and she subsequently became the much-loved pet of his daughter Kathy. In more than thirty-five years of rabbit breeding this was the only one eared rabbit that James had ever observed.

In autumn 1984, Unicorn escaped from her pen, but three days later she was found,recaptured, and placed in a new hutch. Over the next month, she grew steadily fatter, and 31 days after her original escape Unicorn gave birth to a litter of five offspring. As she had never been introduced to any of the farm rabbits,it is clear therefore, that during her brief period of freedom Unicorn had encountered and mated with a wild rabbit.

Of her five offspring, four were normal, but the fifth displayed its mother's remarkable median-ear condition. Regrettably, however, all five offspring died shortly afterwards during a very severe thunderstorm, so no details of their sex are known. Happily, Unicorn survived, and lived for a further two years, but she did not give birth to any further litters, so the unidentified mutant gene presumably responsible

Fig 1 Kathy with Unicorn

for her median ear and that of one of her offspring was lost forever when she died in November 1986.

Judging from the 4:1 normal: mutant ratio of offspring, it is likely that the median-ear condition was induced by a recessive allele (gene form), and that Unicorn was homozygous for it (i.e possessing two copies), thereby enabling the condition to be expressed by her. If so, then it must

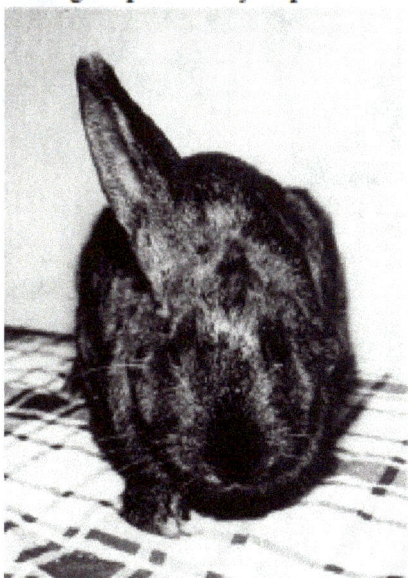

Fig 2 Unicorn the Flemish Giant doe.

Fig 3 (p. 54) Dr Shuker and `The Shaman.`

also be assumed that her wild mate was at least heterozygous (possessing one copy) for this same mutant allele, in order to explain the birth of the single median-eared offspring in her litter. Yet if this mutant allele is indeed present in the wild population, one might have expected it to have been expressed far more frequently (especially in animals that are famous for breeding...well, like rabbits!). Could it, therefore, be associated with some debilitating trait too, so that individuals expressing it are more vulnerable in some way to predation?

The most obvious affliction to be expected that may prove detrimental to survival in the wild is some form of hearing impairment - an occurrence that normally accompanies most ear-related mutations. Yet Kathy had observed that when Unicorn was called, she would turn towards the direction of the voice, thus suggesting that her hearing was not severely impeded (although by having only one ear, it meant - inevitably - that Unicorn's hearing could only monoaural, not stereo).

Tragically, however, in the absence of further litters from Unicorn upon which to base breeding observations, little more can be said of her apparently unique mutation. So it is likely that its identity will remain undiscovered, unless this remarkable 'unicorn ear' condition reappears one day in some other rabbit farm.

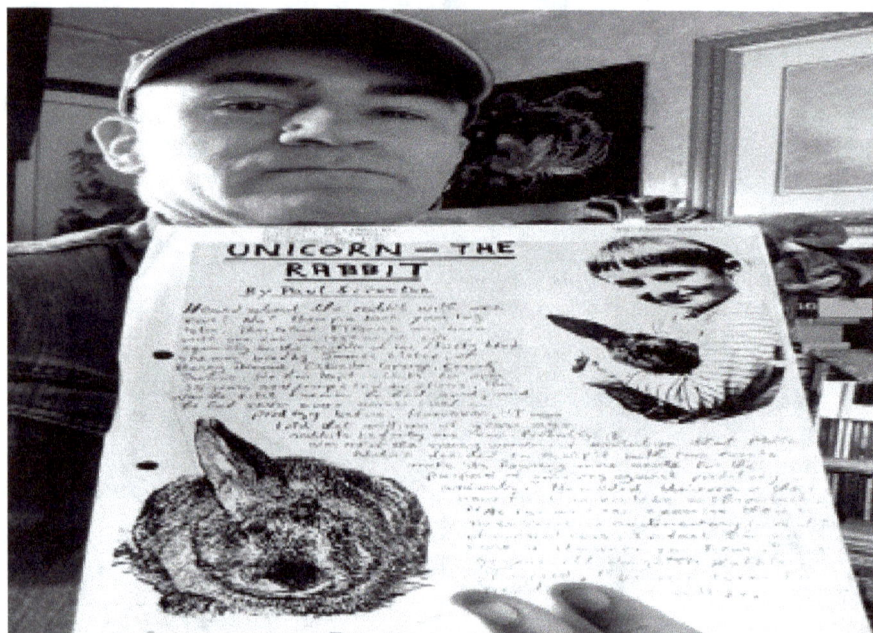

Fig 3 Dr Karl Shuker with a copy of The Shaman.

A Spotted Otter in Ireland

Richard Muirhead

This item relates to a spotted otter from Lough Sheelin, Ireland, c.1909 and my correspondence with Dr Andrew Kitchener of the National Museums of Scotland on the subject. The original story is to be found in the Irish Naturalist vol XVIII 1909, pp 141-142 by R.F.Scharff . 'On the occurrence of a speckled otter in Ireland.' "The National Museum of Ireland recently acquired from Mr W.J. Williams , of Dublin, a full- grown Otter, which differs from ordinary otters, in that its rich brown fur is spotted all over with white spots......It was trapped in Lough Sheelin, which lies partly in the County Cavan and partly in Westmeath.

The fur, as a rule, is of a rich chestnut brown in Irish Otters. It is composed of two parts, the woolly under-fur and the longer stiffer hairs projecting beyond this. The colour of the under-fur is very light grey changing to greyish-brown above, while the longer hairs are chestnut brown throughout. On closely examining an Otter skin we sometimes find that the under fur is not quite uniformly coloured throughout. Here and there,separated by wide spaces, very small perfectly white tufts are met with in the under-fur.When these occur in such large patches as to be clearly noticeable, their presence impairs the value of the skin from a furrier's point of view. Mr Williams informs me that occasionally, amounting to about 1 per cent of the skins are speckled in this manner. The whiteness, however, in these cases, is still hidden to some extent, in the unprepared skin, by the brown colour of the long hairs. It is only after removal of the hairs by the furrier that the white spots become plainly visible.

In the specimen here figured, not only has the under-fur white patches of variable size, but the whiteness extends even to the long hairs, giving the Otter a most peculiar speckled appearance. Mr Williams tells me that, among several thousands of skins that have passed through his hands, this is the only specimen of that kind he has seen.

From the Royal Irish Academy Fauna and Flora Committee's records, I find that perfectly white Otters have been observed in the River Shannon, being,presumably true albinos, and recorded in the *Field* (vol. xci., 1898, pp.141-42). We know that an albino Otter from Scotland is preserved in the Belfast Museum, but the only record of a speckled Otter that I have noticed is of one supposed to be in the Hancock Museum in Newcastle-on-Tyne. On applying to Mr E.L. Gill, the Curator of this Museum, about this specimen, he kindly informs me that there is no such otter in the Hancock Museum. The Irish speckled Otter now described is therefore, as far as I can ascertain, a unique specimen.

In connection with this very abnormal skin of the Otter, I re-examined the ordinary ones with a view to verifying Mr.Ogiliby's statement (1) that Irish Otters differ so much from English ones as to deserve a special name. He proposed to call the Irish Otter *Lutra roensis* instead of *Lutra vulgaris* .

As there are no English skins of the Otter in the Irish National Museum, I was unable to compare the two externally. Mr.Ogilby gives the dark colour, which he describes as almost black, as the principal character of the Irish Otter. There are about half-a-dozen Irish skins in the National Museum, but none of them are darker than chestnut brown. Mr.Ogilby mentions that there are differences also in the size of the ears and in the proportions of other parts in the Otters from the two countries without, however,indicating to us anything more definite. More recently Dr. Scalter has drawn attention to this alleged difference in the fur, while William Thompson (2) thought that the skull of the Irish Otter was larger than that of the British, and stated that Dr Robert Ball considered that two Otters to be, perhaps, distinct varieties.

None of these authorities clearly define how an Irish can be discriminated from a British Otter. There are sixteen recent Otter skulls from various British and Irish localities in the National Museum. After a careful examination of these I failed to detect any character by which Irish skulls can be distinguished from British ones. At any rate we cannot argue from a comparison of a few skulls that the Irish Otter is larger than the British. The size of a species varies in every country between certain very definite limits, and, as the largest specimens are generally sent to a museum for preservation, a large series is required to determine the average size. As far as the skull is concerned, I think the Irish Otter is not, specifically, distinguishable from the British." (3)

On 29th April 1996 Dr Andrew Kitchener wrote to me saying :

Dear Mr Muirhead, Thank you for your letter of 26th February. I am sorry that I have not been able to answer until now, but I have been tied up with various urgent projects. The otter looks most odd. I think there are two possible explanations. Either it is a hoax, or it could be a simple single gene mutation causing white spotting. This white spotting gene is known in cats and there is no reason to suppose it does not occur in otters, although as in most wild mammals, these gene mutations are rare. Anyway, thank you for sending me the photocopy- it is a most unusual otter indeed. Yours sincerely, Dr Andrew Kitchener.(4)

REFERENCES

1. W.Ogilby Proceedings of the Zoological Society of London, 1834 (part 2) pp 110-111.

2. W.Thompson Natural History of Ireland, vol 4 1856 p.6.

3.R.F Scharff   Irish Naturalist vol XVIII 1909 pp141-142.

4. Letter from Dr A.Kitchener to R.Muirhead April 29th 1996.

Fig 1 The Spotted Irish Otter of Lough Sheelin.

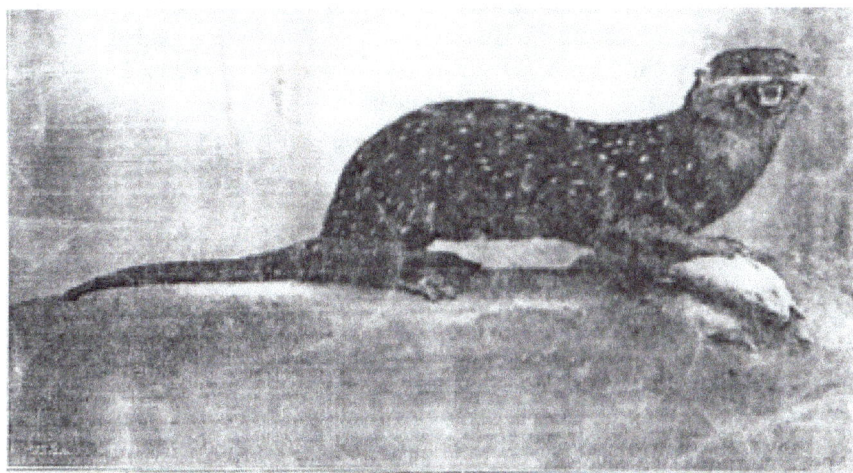

Fig 2
Approximate
Location of
Otter

Reactions to
Flying Snake 1

Karl Shuker kindly gave me permission to reproduce his Shuker Nature blog of April 16th 2011 which commented as follows on Flying Snake 1:

" It`s always great to see a new cryptozoological periodical, especially one in hard-copy form, with pages that you can touch and turn with your fingers instead of with an impersonal, intangible tap of a stylus - call me old fashioned, but for all their promise of instant access, e-books to me are nothing more than soulless text, whereas a real book is also an experience, even a friend, to enjoy and rejoice in.So I welcome with unadulterated enthusiasm the long-awaited, much anticipated first issue (April 2011) of *Flying Snake* - fellow cryptozoological investigator Richard Muirhead`s brand-new hardcopy journal of cryptozoology,folklore and forteana." (Having said that, Richard does plan to publish a pdf version in due course, but he will continue with the hard-copy version too)

"Like me, Richard has a particular interest in unearthing very unusual and obscure reports from the literature and in collecting hitherto unpublicised accounts from correspondents, and Flying Snake certainly does not disappoint. Within its professionally-produced 68 pages, it covers a veritable crypto-cornucopia of extraordinary subjects from around the globe - including reports of flying snakes (naturally!) from Wales, as well as unidentified flying lizards in Australia, pinked tusked elephants from China, the devil crabs of South Shields, a mermaid from Israel, a couple of fascinating Nandi bear reports that were new to me, giant centipedes in Hong Kong, an article by me concerning a previously obscure equine cryptid from Iberia, and lots more! "

With a planned publication schedule of 3 issues per year, at a cost of just £3 per issue or £9 per annum, *Flying Snake* promises to be a very worthy investment for anyone interested in cryptozoology and wider animal-related mysteries or anomalies.

Marco Masseti commented thus on Flying Snake 1 in an e-mail to me of April 23rd 2011:" Dear Richard, I just received the copy of the first issue of *Flying Snake*. Compliments, this is a very amazing and interesting publication."

All the best Marco.

Matt Bille, one-time editor of Exotic Zoology wrote the following on his web site Matt's Sci/Tech Blog http://mattbille.blogspot.com/ May 25th 2011

Neat new cryptozoology magazine

"Richard Muirhead sent me the first issue of his new " Journal of Cryptozoology, Folklore, and Forteana," *Flying Snake* . It's not a journal in the peer-reviewed sense, but a very enjoyable little magazine. The printing is crisp and professional, the articles generally well written, and the correspondence included from Richard's files very interesting. It strays to the borders of cryptozoology and beyond (cf. Richard's own piece insisting the Biblical story of Ezekiel's wheels was a supernatural event and not a UFO), but it's Richard's magazine and he can address anything he wants. A notable feature is that not one article addresses the "classic monsters" of cryptozoology, and I found that quite refreshing!

I loved the color photo of an orange badger (really!) on the back cover."

Fortean Times October 2011 no. 280 p. 65 featured a pleasing review as follows:

"FLYING SNAKE New periodicals within our broad remit are as rare as sightings of flying snakes......Oh look! Here's one! Let's welcome this new journal dedicated to the intersection of forteana with cryptozoology and folklore, edited and published by Richard Muirhead, a veteran of these subjects. The first issue carries articles on pink tusked elephants in China, giant centipedes in Hong Kong, a Dorset wild cat, an Israeli mermaid, the Nandi bear, flying lizards and Ezekiel's Chariot (but this seems to be a flyer for a ` Christian` ufology site!")

61

Notes and Queries

Richard Muirhead would like to know whether anyone has heard of a **phobia of butterflies**.

I would also like to hear about the story of Richard 1 , Richard the Lionheart (reigned 1189-1199) introducing the **Mute Swan** from Cyprus to England.

The image on page 49 of a rabbit with tusks prompted me to do a bit of research on the old Net and this threw up as many questions as the original **rabbit with tusks** story .Particularly, from the `Jackalope Cousins?`(1) web site. Please can anyone express an opinion as to the identity of the following?:(I am placing these stories in date order though on the web site the last story appears first.)

In 1572, the Corando (2) expedition searching for the fabled Seven Cities of Cibola, recorded seeing a `lion-rabbit`pouncing upon its prey from a low tree. This was discounted as the expedition recorded numerous new species that were unfamiliar to the Old World explorers.

In 1878, a Tombstone Arizona prospector known only as `Pete`, arrived in town with stories of being attacked by a creature that moved too swiftly to be identified. Pete appeared badly scratched with two puncture marks just under the knee. Quills were also found in Pete's clothing giving evidence of a struggle with the creature.

From an unidentified source by e-mail:

Officials and leading zoologists are baffled over a 1996 sighting of a previously thought of legendary animal. The creature, locally called a `Razor-Jack`, was reportedly sighted in an area of the McDowell Mountains,north-east of Phoenix Arizona.

The Sabre-tooth Bristled Hare was described as looking just like the common jackrabbit but with two distinguishing features;tusks and a patch of quills(like a porcupine's) on the animal's back. The tusks are apparently similar to the exten-

-ded canine teeth of the region's Javelina or collared Peccary. Interestingly also is the fact that a Javelina's coat is not fur but is made of hollow quills. This startling implication has officials putting a tight lid on this story. While unusual,this sighting only reinforces other sightings recorded in history that have made 'Razor-Jack' a legend."(3)

REFERENCES

1. Jackalope Cousins? http://users.stargate.net/~mnovak/jackalopes/cousins.htm

2. This expedition actually took place between 1540-1542 in N.Mexico and the S.W U.S.A.

3 Jackalope Cousins ? web site op cit.

A **homing snake**: I was told by a friend that he had read in the Macclesfield Express in around 2001 about a snake that had been trained to find its way home to its owner from one part of Macclesfield to another,a distance of about 2 miles across busy roads and parks. Is such a thing possible and has it been heard of before?

Some notes from John Aubrey's **Natural History of Wiltshire** (written between 1656 and 1691) might be of interest:

"In Cranborn Chase and at Vernditch are some martens still remaining...In Wiley river are otters,and perhaps in others. The otter is our English bever;and Mr Meredith Lloyd saies that in the river Tivy in Carmarthenshire there were real bevers heretofore- now extinct. Dr Powell, in his History of Wales,speakes of it. They are both alike;fine furred, and their tayles like a fish. (The otter hath a hairy round tail, not like the beavers-J.RAY).....In warrens are found,but rarely,some old stotes,quite white:that is, they are ermins."(1)"In Sir James Long's parke at Draycot-Cerne are grey lizards;and no question in other places if they were look't after; but people take them for newts....At Neston Parke (Col.W.Eire's) in Cosham Parish are huge snakes,an ell [1] long; and about Devizes snakes do abound.Mrs Fr.Tyndale,of Priorie St.Maries,when a child,voyded a lumbricus biceps. Mr Winceslaus Hollar,when he was at Mechlin,saw an amphisbæna,[2] which he did

---

[1] That is, 45 inches.

[2] See Richard Muirhead 'The Amphisbæna in Britain and Ireland' CFZ Yearbook 2011'

very curiously delineate, and coloured it in water colours, of the very colour: it was exactly the colour of the inner peele of an onyon: it was about six inches long, but in its repture it made the figure of a semicircle;both the heads advancing equally. It was found under a piece of old timber, about 1661; under the jawes it had barbes like a barbel, which did strengthen his motion in running. This draught, amongst a world of others, Mr. Thom.Chaffinch,of Whitehall,hath; for which Mr. Hollar protested to me he had no compensation. The diameter was

about that of a slo-worme; and I guess it was an amphisbænal slo-worme.[3]" ["The serpents called amphisbæna are so designated....in consequence of their ability to move backwards as well as forwards. The head and tail of the amphisbæna are very similar in form: whence the common belief that it possesses a head at each extremity. It was formerly supposed that cutting off one its "heads" would fail to destroy this animal; and that its flesh, dried and pulverised, was an infallible remedy for dislocations and broken bones-J.B." (2)]

REFERENCES   1. Aubrey's Natural History of Wiltshire p.50   2. Ibid p. 74

John Aubrey (1626-1697)

Wikipedia Creative Commons

---

[3] How I would dearly like to find out if this illustration survives somewhere.I contacted the Ashmolean Museum in Oxford but that lead went nowhere. Richard.

BOOK REVIEWS

Varmints Chad Arment Coachwhip Publications 2010 (ISBN 1-61646-019-9 ISBN-13 978-1-61646-019-8)

This book is so good a law should be passed, backed up by the United Nations or U.S Government saying it ought to be on your shelf,or that of your friends whether they are cryptozoologists or not. If Chad intended his all 682 pages of his book to be a kind of encyclopaedia of N.American cryptid carnivores then he certainly achieved his goal. He humbly states however that it is a "preliminary investigation." I hope Chad is planning similar books to go alongside Varmints and his earlier Boss Snakes for other members of the animal kingdom, say marine life, birds etc? Varmints is divided into the following chapters: What accounts for Varmints?, Native Carnivores, Exotic Carnivores, Varmint Folklore, Varmints by State and Province, Evidentiary Requirements. Appendix A: Predator Kill Patterns, Appendix B: Basic Profiles and Tracks and Bibliography. There are also black and white illustrations of the animals mentioned and occasionally one of the cryptids included e.g the giant otter caught in N.Maine in 1949 on p.82 and the equally extraordinary Euroopean Wild Cat, killed in N. Pennsylvania, 1922, on pp 542-543, which Chad makes clear from the newspaper extracts he includes, may not have been as scarce in N.America as first thought.

Chad explains that "Varmints" is a name given to " the first folkloric step in creating the perception of a distinctive animal , separating it from the commonality of known predators and elevating it to ethnoknown status, even if it later turns out to be mundane." (p.9) Even within well known species such as the bear there are species newly or fairly recently discovered such as the lava bear or dwarf grizzly bear of south-eastern Oregon and possibly adjacent Idaho. The whole picture is complicated by a plethora or misunderstandings and invented words (e.g "Dwayyo," "Glawakus", etc.)

The vast majority of the book is composed of historical reports from newspapers from Alamaba through to the Yukon Territory, by state and province, from the 19th century through to the present day and not just the obvious critters such as panthers and lynxes but also less likely candidates such as wolverines, margays and also coatis occurring beyond their normal range. It needs to be pointed out that the author deliberately excludes the eastern cougar " which technically is not cryptozoological. " (p.18) There is a thorough bibliography.

My only minor recommendation is that due to the great length and weight of Chad's book it might have been better to split it into two or three volumes. Dr Devo's major recommendation, rather decree, is don't be a 'Blockhead' (N.B.Devo fans) - buy it!

Big Cats Loose in Britain Marcus Matthews CFZ Press 2007 (ISBN 978-1905723-12-6)

Here is another substantial work of cryptozoological erudition, though somewhat less deep in its scope in comparison to Varmints. In Part 1 there is an introduction and coverage of S.W Britain's mystery cats. Part 2 covers the Surrey Puma and other S.E big cats, Part 3 covers the rest of the country. Marcus is particularly strong in his coverage of the Exmoor beasts and Surrey pumas, spending 45 worthwhile pages on the former and 65 on the latter. I did not know that the Exmoor beasts go back as far as the 1870s having become aware of them in 1983. However there is a big gap in Surrey puma sightings between the possible one by William Cobbett's in 1770 and the next probable cases in the 1950s, undermining Marcus's case elsewhere in the book that big cats in Britain may originate as hybrids from Roman times. The Romans army withdrew from Britain to defend Rome c. 410AD

The whole Exmoor beasts saga is far more complicated than the British media would have us believe, judging by Marcus research.[Di Francis research is mentioned with Marcus wisely reminding the reader that her theories " proved too much for some learned scientists" (p.14)] Not only has there been a possible unique species of British big cat (see Francis) in Devon and Somerset but even lynxes and a werewolf are claimed as candidates! (the werewolves were on Exmoor and Dartmoor). Marcus is less strong in Part 3 of the book spending less time on the Isle of Wight mystery cats, Welsh mystery cats, mystery cats in the Midland and the Upper Shires and Northern mystery cats. More space could have been spent on the Appendix: Other Mystery Animals of Britain but I concede it's impossible to trace every single anomaly/escapee.

Kraken 3   Dépt de cryptozoologie Bernard Heuvelmans   Musée de zoologie Lausanne Mai 2011 (ISSN 1662-4696)

Kraken is the occasional publication of the Dépt de cryptozoologie Bernard Heuvelmans but it is well worth the wait. Unfortunately for me,more than half of issue 3 is in French for obvious reasons. I cannot read French. Issue 3 strongly concentrates on mystery hominids. The English language essays are:-

Gustave Sànchez Romero.

Olitiau and Kongamato: African winged reptiles: The fabulous pterodactyls of the black continent.   33-47

Michael A.Woodley.

Introducing Aequivotaxa: A new classification system for cryptozoology. 63-85

Both these are very interesting and worthwhile in their own particular way, though I found Woodley`s difficult. Romero`s particularly appealed to me because of my own interested in Namibia`s flying snake. It is good to know Heuvelman`s archives are getting a wider airing despite the continuing absence of many of his books in English. However hopefully Kraken will grow in popularity thus increasing the likelihood that more of Heuvelman`s research will be revealed. I have a feeling this will happen.

The French articles are as follows: -

Benoit Grison.

Psychologie soviétique, histoire sociale et anthropogenese la " lutte pour les troglodytes" de Boris Porchnev   3-19

Annette Jubara

La recherche de l` origine de l`homme chez Boris Porchnev   21-31

Florent Barrere

L`imaginaire du Poulpe colossal   49-61

Yannis Deliyannis

Jean Céard, La nature et les prodiges: L`insolite au XVI siecle.   89-92

Letters to Flying Snake

Here is an old letter from Dr Sigrid Schmidt, an expert on the Flying Snake of Namibia:

Dr. Sigrid Schmidt   Hildesheim, 16.9.95

Am Neuen Teiche 5

D-31139 Hildesheim

Dear Mr Muirhead,

Thank you very much for your letter which I received this week.
Certainly I can give you some information on the strange Namibian
snakes. For they belong to the foremost topics of Nama legends and
folkbelief , a field in which I have specialized.

Like the ghosts or UFOs in Europe, these snakes are seen by people who
believe in them. And people who do not believe in them do not see them.
A teacher once sarcastically told me: If at night people see the light of a
motor-bike or a car where only only one light is working people say:Oh,
there is the snake again!And there are many people who delight to tell
tales how they saw the snake or about other people who met the snake.
Usually quite a number of different traits are attributed to these snakes,
each narrator stresses different ones: its stench which alone kills people
and attracts swarms of flies, its call which sounds like sheep or goats
calling, the light, lamp,mirror, stone or white spot in the forehead, its
face like a man`s face, sometimes even with a beard, its horns or ears,

its fondness for women. In dry Namibia the snake (which is usually called the big Snake) lives in the mountains, but in the permanent rivers, particularly the Oranje River, its aquarian equivalent lives in the water, has a palace under the water and keeps there his human wives which he steals at the shore. These snakes belong to the very ancient belief in Africa and in other continents as well. In southern Africa there are rock paintings of prehistoric times of huge snakes which probably were connected to rain or rain ceremonies.

As to the flying snake in particular: Usually this snake has no wings but uses the end of its tail to push itself through the air to the next point. And as to the reporter of the 1942 accounts: the policeman Honeyborne was known as a very good narrator and experienced quite a number of extraordinary things.

I published a few tales about such mythical snakes in my book "Marchen aus Namibia" (Koln:Diederichs,1980), pp.232-236, and I shall devote a chapter on it in the book on Namibian legends which I am working now.

I hope this brief information will help you a little. If you need more, please, just write me. But the field of African mythical snakes is endless because of the many ancient connections to religion and ceremony.

Yours sincerely, S.Schmidt

The following e-mail came from Irene Brierton of the Mid Derbyshire Badger Group on March 29th 2011. The "erythrisitic badger specimen" she refers to is the one on the back cover of Flying Snake 1.

"Dear Richard

Your orange coloured badger was an example of a variant in colour known as erythristic, much the same sort of thing as black,grey squirrels,which occur in some parts of Britain, described as melanistic. The colour results from the presence of a recessive gene, present in the genetic make up within some social groups of badgers.

Being a recessive gene, an erythristic animal may breed its whole life through producing only normal coloured young, however the gene will be passed on resulting,sometimes generations later,in another similarly coloured badger. This colour can vary in individuals from pale fawn, through distinct ginger,to more rarely a quite dark reddish brown colour.

This gene is present within some social groups of animals in Derbyshire, as it is in other parts of the country, but I understand there are areas where it is much less evident,if not absent altogether.

The High Peak Badger Group deal with badger issues in the far north of the county and your badger may belong to them. I cannot remember if they have an erythristic specimen and if they do I am pretty sure they will not be prepared to sell it to anyone.They can be contacted on 01298 26957.The Mid Derbyshire Badger Group have an erythristic badger specimen which we take on talks, along with a couple of normal called individuals. Ours originated as an unfortunate road casualty we had been informed about and is not for sale.Although we do occasionally receive reports of these dead animals being found,by far the majority of such reports are of normal coloured badgers. We do not save them as a matter of course and they are usually disposed of in the usual way.

Regards

Irene Brierton

Below is an extract from an e-mail to me of April 13th 2010 from John Didier, Associate Dean, College of Liberal Arts and Associate Professor,Department of History Colorado State University concerning the black pink tusked elephants of T`ang dynasty China which I covered in Flying Snake 1. I had tucked away this e-mail in a file somewhere and fortunately found it. The first paragraph covers what Schafer had to say in 'The Vermillion Bird' about the elephant. In the following two paragraphs Didier provides more information as follows:

".....I've also checked the Chinese text of the 16th-century "Chinese Materia Medica" (Bencao gangmu), the standard (and enormous) encyclopaedia of these types of zoological,botanical,medicinal,etc. notices in/from/Chinese history/civilization,but while the treatment of elephant is fairly lengthy,there is no mention of these black elephants with pink tusks from Nam Viet.Probably,then,there won`t be much to find.

Schafer doesn`t actually say anything that I have found about the extinction (timing or otherwise) of this type of elephant, and I`m not certain from where the websites that you referenced in your email might have obtained their estimate/report that these elephants went extinct in the 14th century. I would think that either report would be questionable in its accuracy-in that world in which most of Vietnam was still undisturbed forested mountain territory inhabited by virtually or totally undisturbed forested mountain territory inhabited by virtually or totally undisturbed Neolithic tribes,how would one really know when/if these elephants had become extinct.

GIANT TURTLE CAUGHT
NEAR BRISBANE

A YEREN IN A MACCLESFIELD

ANTIQUE SHOP?

The photograph below shows a statue of a wild man

In Aladdin`s Cave, an antiques shop in my home town.

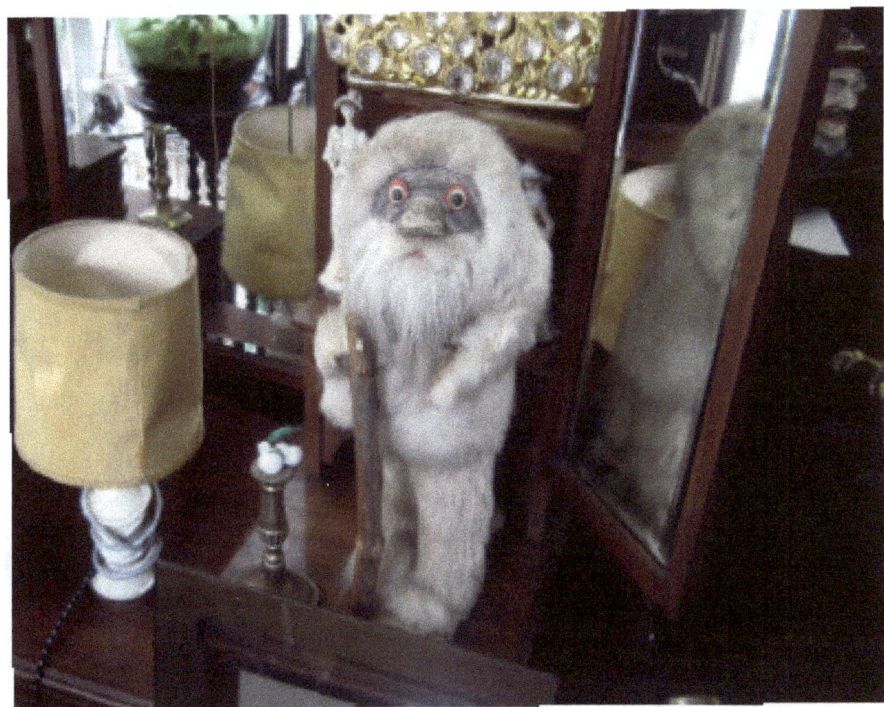

Image on page 70 from The Mercury, Hobart, January 14th 1947. Thanks to Trove and National Library of Australia.

This poster, left, alerted Chinese people to look out for the Yeren. There is a distinct similarity to the model above. Thanks to Loren Coleman for permission to reproduce.

Flying Snake

A Journal of
Cryptozoology, Folklore and Forteana

Volume 1   Issue 3   May 2012   £3

ABOUT FLYING SNAKE

Flying Snake is available from:
Richard Muirhead
Flying Snake Press,
112 High St,
Macclesfield,
Cheshire,
SK11 7QQ
UK
http://homepage.ntlworld.com/richmuirhead/cryptozoology/

Tel: 01625 869048

Mike Hardcastle,Sub-Editor,NSW Australia .   Zvi Ron Israel correspondent.
www.steampunknaturalist.com   Carl Marshall   Zoological Consultant.

Please feel free to contact me if you want to reproduce anything I have written. If you want to reproduce other authors' works, I will try and contact them on your behalf and get back to you. The opinions of authors other than myself do not necessarily reflect my own.

Back issue available on request.

PAYMENT

Subscriptions: £ 3 per issue, £ 9 per annum. PDF via e-mail. £3

Payment for however many issues you (and your friendly neighbourhood, reading, flying, snake) would like to purchase can be made by means of PayPal on my web site (See url above).Or via http://www.flyingsnakepress.co.uk

Cheques and postal orders from within the United Kingdom should be made out to Richard Muirhead, and *not* Flying Snake.
Cheques will not be accepted from outside of the U.K at this point in time.

CREDITS

Thanks to Carl Marshall for the photo on the cover of an Arctic fox and Mike for the drawings in the China and Hong Kong cryptozoology essay. Also,thanks to Mailbox, Macclesfield,again,for printing.

CONTENTS

DR DEVOS DIARY

"For I pray God for the introduction of new creatures into this island. For I pray God for the ostriches of Salisbury Plain, the beavers of the Medway and silver fish of Thames." Christopher Smart, `Rejoice in the Lamb`. Poet, naturalist, lunatic (1722-1771)

For a long time this Truth has been suppressed, but now it can be revealed to every man, woman and mutant on the face of this planet - issue 3 of Flying Snake has arrived on your doorstep (with apologies to Devo`s General Boy for that quote) or inbox if you`re one of those select few (er, select one actually, Mike in Australia) who receive it by e-mail.

Talking of electronic media and such like, Flying Snake will soon be available on Kindle if I get my act together, I thank Vikki Barlow my manager at Oxfam, Macclesfield for telling me about this possibility. I am quite excited about this, being aware that finances plus benefits help the chances of not starving to death and keeping a roof over my head and the opportunity of doing more crypto research. Which brings me on to the developments since Flying Snake 2 was published last October. I was hoping to bring you an account of an eye witness observation of the Loch Ness Monster from the early 1970s, but because I went on about it to the person who saw it like the crazy loon I sometimes am, (as in lunatic, not the bird of that species) a bit to much, she changed her mind. But as I barely believe in Nessie and this is a magazine about obscurer cryptozoology, perhaps it`s for the best?

I have tried to include the subjects mentioned in the back of FS2 in this number and it is more Fortean than usual, including an appearance of the Virgin Mary and baby Jesus on the roof of a caravan. I love the way this subverts the usual expected locations, e.g roof of a grandiose cathedral or monastery.

I would appreciate your feedback, now that Flying Snake is one year old with this issue and I`m sorry it`s late. I hope to include telling the time from cat`s eyes, the tragic story of Ota Benga, the man caged in a zoo and part two of Carl`s Giant Snakes essay in Flying Snake 4. Enjoy!

Fortean Foxes

Richard Muirhead

By "Fortean Foxes" I mean foxes seen in Britain of odd coloration size,or behaviour. Or any other Fortean aspects. I have on record a blue fox from about the 1870s or `80s. This story is from a book about hunting in Co. Durham, Britain:

" Foxes vary in colour very much,and we killed a blue one in the West country one season about this time, [November-R] after finding him at Mainsforth whin." (1)

According to Karl Shuker: " The blue fox is bred for its fur so perhaps there were fur farms there at that time?" (2)

Arctic foxes occasionally turn up in Britain. The one on the front cover was killed in Nuneaton, Warwickshire. The photo was kindly provided to me by Warwickshire cryptozoologist Carl Marshall who said that it was killed by a dog (probably a Jack Russell terrier) in the late 1970`s .

In April 1983 an An Arctic Fox (Alopex Lagopus) turned up in Dorset:

" A strange light-coloured fox was seen alive on the farm of Mr J. Strange at Worth Matravers (SY / 9776) near Poole, Dorset on a few occasions between about 4th and 15th April 1983. On 16th April the animal was found recently dead in a field one kilometre away from the nearest road. The body was given to one of us for identification and autopsy to ascertain the cause of death… Two colour phases exist;one of which turns completely white in winter. This particular animal was pale fawn and grey in colour with dark tips to some guard hairs and the face was very dark grey. The pelage has long guard hairs with dense underfur. The large paws have much fur between the pads. The ears are noticeably small and rounded .The appearance is of a small and very long haired, light-coloured fox with relatively smaller ears, shorter muzzle and shorter legs than the Red Fox (Vulpes vulpes) .

Apart from the ears some of this appears to be an illusion due to the long hair as in this particular animal, relative to head and body length, the dimensions were not greatly different to that of the Red Fox... However there is always the possibility of it being a hybrid. A. lagopus is adapted to arctic conditions... Obviously the animal has arrived from the arctic under its own power suggesting escape from a fur farm or zoological park. It had managed to survive successfully and feed for at least 10 days as shown by sightings and the faeces in the rectum. It was not thin and there was a considerable amount of fat in the omentum. It would be interesting to know from where the fox escaped or was released and so the distance travelled and the length of time it was able to survive at liberty". (3)

The April 2010 issue of Fortean Times contained the story of a platinum fox seen on Dartmoor. Karl Shuker wrote:

"On the evening of 13 March 2010 , Shaun Histed-Todd was driving a bus along a Dartmoor road when he saw a most unusual creature run down the edge of the moor and stand by the roadside , where the buses headlights afforded him an excellent view of it for half a minute before it ran back up onto the moors... Shaun contacted me a few days later, as he was unable to identify it, and provided me with a detailed description. It resembled a young fox and had a bushy white-tipped tail but its coat was dark-silvery-grey; it had noticeably large ears, white paws and a black raccoon-like facial mask. Reading this, I was startled to realise that Shaun's description was an exact verbal portrait of a most unusual yet highly distinctive animal - a young platinum fox. After checking photos of platinum foxes online, Shaun confirmed that this is indeed what he had seen..." (4)

Certain individuals with, in Fort's words ," wild talents" seem to be able to "fascinate" foxes. The late William Corliss recorded an instance from India in his `Incredible Life: A Handbook of Biological Mysteries` (p.265). Citing English Mechanic 62: 269-70 1895.

"A man began a muffled, chuckling kind of call, which he kept up without ceasing. In about two minutes a fox came out of a little ravine close by, and,looking suspiciously about him, trotted towards the noise."

The article goes on to say how eventually 30-40 foxes gathered. ("The instant the man stopped his chant, every one of the animals fled, as if the spell was broken.") (5)

More obviously Fortean behaviour includes:

"T.A. Shellswell wrote to *Reville* (20 Feb 1976) that a fox being chased by a local hunt once sought sanctuary in a neglected corner of his garden, later escaping safely. Every year since , wild fox-gloves have grown there, where nothing grew before." (6)

"Also from *Reville* (9 June 1978) we learned of a young fox who lost his way in Thame, Oxfordshire, and took refuge in the cellar of a pub called The Fox." (7)

"G.F. Tomlinson wrote to the Sunday Express (28 Aug 1981) that he was playing golf with a friend when they noticed a fox basking in the sunshine. His friend played a shot which landed near the animal, and it grabbed the ball and made off with it. Mr Tomlinson gave chase, and the fox dropped the ball. When examined , the ball had R. Fox and Sons" written on it, with a red fox logo."(8)

In July 2002 the Manchester Evening News Media web site reported on an attack by a five legged fox on a man in the Audenshaw area:

DAD KILLS FOX WITH FIVE LEGS

"Dad Jason Lloyd killed a FIVE-legged fox after it attacked him in his kitchen.

Jason was cooking sausages when it crept up behind him. He said " I`d opened the back door to let some air in when suddenly this fox pounced. "It tried to attack me, but I managed to push it away. Then it ran into the garden and my Jack Russell, Pip, started fighting with it.

I thought it was going to kill the dog, and I was worried because my little boy Jack was in was in the house…..I hit the fox with a yard brush and when I looked it was dead. But then I got the shock of my life when I realised it had five legs…..Recently a fox attacked a young child in a house in the south of England." (9)

Perhaps this fox turned aggressive through the trauma of having five legs? I mean, wouldn`t you?

Philip Gosse mentions a case of fox fascination in The Romance of Natural History. Second Series 1861

" Mr J. H. Gurney reports the account of a respectable gamekeepers , who, being much annoyed by the nightly visits of a fox to the poultry, could not imagine how Reynard managed to effect his purpose, as they roosted on a large spreading oak. One morning, however, just as day was dawning, he heard a great noise among the poultry, and, looking out of the window, saw a fox running round and round under the place where they sat, and soon observed the fowls began to fall from the tree in great confusion. The fox immediately seized his victim, and the mystery was so far solved. A day or two afterwards the fox , a very large male, was killed in an adjoining paddock, and no further assaults were made upon the poultry." (10)

On May 31st 2010 the BBC Nature U.K. Web site posted:

"Over 20 years ago I was tipped off about an incredibly rare black fox in southern England and was lucky enough to film it. I have not seen one since...
The Springwatch team have been searching all over the country to see if any are alive today, and hopefully to film one, but unfortunately the trail has gone cold. Although I've heard rumours of sightings here and there, I'm intrigued to know whether or not there are any living examples of these near mythical beasts. And I really need your help.
If you have ANY information about a black fox, please, please let us know by commenting below. There was a confirmed sighting in Lancashire some time ago, but nothing recently as far as we know." (11)

Another oddity, this time from the www.husky-owners.com Forum dated February 4th 2012, a posting by Ron Tao+Sky in Dover, Kent

"The strange thing that is puzzling me is the amount of dead foxes that are on the roads near me. Its not just the odd one but 3 or 4 together as though they have been dumped there. Now it's happening with the Badgers." (12)
Some time in 1995 the late Jan Williams,whilst editor of SCAN (through whom Jon and I "rediscovered" each other after our time together in Hong

Kong) wrote to me about white foxes and other things:

"A pure white fox was killed in 1887 by Taunton Vale Hounds in West Somerset." This is in Man and Beast by Ron Freethy. (13)

In Country Sportsman 1949 (page number unknown) there is a story: `Albino Foxes In Northumberland A Strain That Persists Around Rothbury`

"In the year 1937 a white fox was killed in the grounds of Brinkburn Priory almost within sight of the River Coquet by the Percy Hounds. The mask is now believed to be at Alnwick Castle, the property of the Duke of Northumberland. In the spring of the same year,Richardson, a keeper on the Cragside Estate further west along the Coquet Valley, dug out a white vixen which had two cubs of normal colour. On Tuesday,15th 1938, the Morpeth Foxhounds, whilst hunting in the east part of their country, roused and quickly killed a pure-white fox. The mask, beautifully mounted, is now at Meldon Hall. The tips of the ears and brush are black, the eyes yellowish and lighter than the eyes of a normal fox, and there is none of the pink colouration one associates with the true albino....Later in the season another white fox was reported as being seen by the rabbit catcher at Paxton Dene, but no further trace of it was found. This outcrop of albinism naturally caused a good deal of local interest and there were many wild and fantastic theories as to the cause of this phenomenon. One of the most popular was that these white foxes had all been fathered by a silver dog fox which, by a strange coincidence,had escaped early in the spring of 1937 from a silver-fox farm in the neighbourhood of Capheaton,not very far,as a fox will travel,from the Coquet banks....In 1947 the keeper at Linden Hall, which lies about a mile north of Paxton Dene, reported having seen a white fox in the Dene".

"This man's evidence can be taken as reliable.....The seed of this white breed, I feel sure, originated somewhere amongst the rocky, rhododendron-grown hills above Rothbury and, from time to time ,it keeps cropping up as it is handed down from father to son."(14)

Jumping forwards 40 years to The Mail on Sunday August 30th 1987, `Lair of the little white foxes` by Dr Brendan Quayle, which told of with two white cubs, one of which was shot the other "booted to death"(15)." "According to David Bellamy* the birth of an albino or white-skinned specimen of a wild or even domesticated creature, unless specially bred, is very rare." (16) One fox was shot. The story continued concerning the killing of the other fox: "…It was two teenage sons of a local farmers who killed the other white cub and one of its red brothers from the same litter. "Why did you do it?" I asked them bitterly this week. "Foxes are vermin and we were worried they were going for our geese," they told me." (17) Quayle speculated their white pelts would end up at a taxidermist.

[* I wrote to David Bellamy about this story but received no reply. The above incident was in the Border country between England and Scotland. Perhaps the same location as in the Country Sportsman article?]

According to Roger Burrows in A Complete Study of The Red Fox (1988) :

"There are records of white, presumably albino, foxes from Dartmoor, and at least five records of them from Whaddon Chase."… "Russian authors mention blue and silver foxes as being present in the northern (part?) of the red foxes` range, so not all the blue foxes in northern Europe need necessarily be descended from the feral North American form." (18)

The Wild About Britain Forum on the Net reported an eyewitness sighting of a white fox near the Wirral. This was on May 2nd 2006. 'Lemming' thought it was a dog but later: "'.This morning I saw it streaking across the field in front of my house, grabbed my binoculars and saw it was a white fox. White from head to hind legs then it turned a peachy colour. How lovely!!! Just wanted to share my sightings with you all." (19)
On the same Forum in September 2007 there were some interesting communications about black and other coloured foxes.

On September 22nd John (in Coventry) said: "I have just had a shock. I was looking out of my bedroom window and a scrawny looking Black Fox walked past my Bungalow…I have never seen a Black Fox before and must admit that I didn`t know they could be that colour. It wasn`t a full on black but dark enough to be able to pass for black."(20) C C replied giving instances of a black one in

Maesycmmer near Caerphilly and one on September 21st 2007 "on the fields by Carmarthen Bay" with "a black stripe down back, with a stripe running down shoulder blades." (21) C C saw three silver foxes in the past four years in Carmarthen.

Finally, on September 18th 2008 a newspaper website reported the sighting of a black fox on the outskirts of Chorley, Lancashire: Mr Hehir, from Preston, Lancashire, was walking in a cemetery with a friend when he spotted the animal among the gravestones…..Country villagers traditionally told stories of how the fox was as "black as night, so that it could live in a man`s shadow and never be seen." (22)

The Spring of 2012 saw reports of much larger than normal foxes turning up in Britain. On March 4th 2012 The Sunday Times reported, with photographs, that " Foxes nearly three times the average size have been shot by riflemen, re-igniting the debate about the threat they may pose to pets and humans. Two giant foxes killed in recent weeks, one in Aberdeenshire, have weighed in at more than 30lb, breaking the 26lb record set by a specimen trapped in Kent at the end of 2010…In northwest London…One mother described seeing a fox fleeing a neighbour`s garden with a cat in its jaws…Jonathan Reynolds, senior research scientist at the Game & Wildlife Conservation Trust, said: A 35lb fox would have been unthinkable a few years ago. We don`t know why they`re getting bigger, but one possible explanation is that they are getting better fed in urban areas." (23) Carl Marshall saw a very large fox in Warwickshire in the early Spring of 2012. At the time of writing he was going to take hair samples to show Jon Downes in Devon.

Huge British foxes may actually not be something new after all. The Country Man of Winter 1957 reported how some hunts-men spied " a gigantic grey wolf-like creature loping across the fields" in Sussex. The witnesses thought it was a werewolf or the ghost of some giant animal killed long before.

Fig 1 A rather dog-like (to my mind) fox in a pub in Nomansland, on theWiltshire-Dorset border. I photographed it c. early 1990s

This account brought a reply a few issues later "from a Doris W.Metcalf, of that county, who had also seen large grey wolf-like animals in the area before WW2, but accepted them as a surviving remnant of some ancient wolf-fox cross breeding that once stalked the downs and ancient forest of Anderida where Pevensey now sits. She writes: " I always understood they were the last of an ancient line of hill foxes, though I have found them on the marshes too. I first saw one of these huge grey foxes on a summer afternoon near Jevington. My companion and I thought it was a large Alsatian dog as it crossed the track, taking not the slightest notice of us...Another time, when the hounds were in the wooded grounds of Glenleigh Manor, one of these huge creatures suddenly appeared walking along the drive a yard or two in front of us." (24)

KNIGHTWICK: WHAT`S HALF-FOX, HALF WILD-BOAR?

CLOSE ENCOUNTER WITH A MYSTERY

"A mystery animal- half fox cub, half wild boar - is at large in the Worcestershire countryside. The creature calmy walked out in front of a car on Monday afternoon, startling its occupants , Jo Morris, from Suckley, near Worcester, and her mother, Rachel.

At 4pm , they were halfway up Ankerdine Hill at Knightwick, when the animal appeared. " It was a cross between a wild boar and a fox cub", said Jo, an equestrian trainer. " I have lived in the countryside for 30 years and have never seen anything like it, wild or tame. It was the size of a half-grown fox cub with a long Roman nose almost like a wild boar`s , smallish ears and its skin was brown and mottled. " It had a hunched back and I likened it to a hyena as it had a smaller front than back. It had a long thin tail, which ruled it out as being a fox cub, which develops quite a brush early on. "...After some research the mother and daughter dismissed the idea that the animal could have been a coypu, which is native to South America, although groups of them live in areas of Norfolk. They escaped from zoos and now breed successfully in the Fens". (25)

R E F E R E N C E S

1. Author of Fox Hunters` " Vade Mecum" The Sedgefield County in the Seventies & Eighties, with the Reminiscences of a First Whipper -in p.189

2. E-mail from Karl Shuker to Richard Muirhead April 18th 2011

3. Proceedings of the Dorset Natural History and Archaeological Society vol. 105 1984 pp 177-8 An Arctic Fox (*Alopex Lagopus*) in Dorset D.J. Jefferies and R.E. Stebbings.

4 Fortean Times April 2010 .

5 William Corliss Incredible Life p. 265

6 Fortean Times 40 p. 26

7 Ibid p. 27

8 Ibid p. 27

9 http://menmedia.co.uk

10 P.H. Gosse The Romance of Natural History. Second Series 1861 p.258

11 BBC Nature UK May 31st 2010

12   www.husky-owners.com February 4th 2012

13 Letter from Jan Williams to Richard Muirhead 1995

14 Country Sportsman 1949

15. The Mail on Sunday August 30th 1987 `Lair of the Little White  Foxes`
    p. 13

16 Ibid p.13

17 Ibid p. 13

18 Roger Burrows A Complete Study of the Red Fox   Ibid p.p 71,73

19 www.wildaboutbritain.co.uk   May 2nd 2006

20 Ibid September 22nd 2007

21 Ibid September 22nd 2007

22 www.telegraph.co.uk   September 18th 2008

23 The Sunday Times March 4th 2012

24 Country Man Winter 1957 p. 635

25 www.worcesternews.co.uk   August 1st 2003

TWO TYPES OF HEDGEHOG

An image from an old natural history book showing a hedgehog without bristles and one with them.

Archive China and Hong Kong Cryptozoology

Richard Muirhead

ODD BIRD

The Straits Times in July 1887 reported a story of an odd bird in China:

" The *Hupao* says that on the 15th ult , some persons from Peking brought an unheard of animal to Canton, which has a red head resembling a turkey , with green feathers, and its body is like that of a goat, with black wool. It has no tongue, and its food consists of bananas, which it swallows without chewing. This strange creature is exhibited at the yamen of the Provincial Judge (at Canton,) and ten cash are charged to see it." (1)

There is another story of an strange bird in Kwangtung (Guangdong) from 1893, this time a bird with a " head of a human being":

ANOTHER STRANGE BIRD

"A Man hailing from Nanhai district in Kwangtung states that an uncommon looking bird was captured there in the country last winter. The bird stands 3 feet high from the ground, and has the head of a human being, whilst its body is covered with hair several inches in length. On its appearance the country people were much alarmed and a large crowd set out to exterminate the uncommon animal, for fear it should be a dangerous customer to harbour. The person who managed to save it from the mob and had the courage to capture it has made a good thing out of it. In addition to its strange appearance the bird is said to be " a most modest creature" , avoiding the gaze of curious people, as

Fig 1 China's Bird with a "Human Head" Mike Hardcastle

if it were too bashful to let people see it, and it is only at the command of its master that it will show itself." (2)

I can only guess (and it is a pure guess) that this was a juvenile ostrich,(because an ostrich has a very vaguely (to me) human looking head.)

Perhaps this was an ostrich which had somehow found its way ashore in Kwangtung province after one of the "treasure ships" of Admiral Zheng He had been ship wrecked on a voyage back from the east coast of Africa in the early 1420s or before ? However Mike has chosen it to look more raptor-like, (see previous page.)

Or may be it was an escaped pet or zoo exhibit.? According to Menzies:

" The treasure ships returned to China with all manner of exotic items: ` dragon saliva [ambergris],incense and golden amber` and `lions, gold spotted leopards and camel-birds [ostriches] which are six or seven feet tall` from Africa; gold cloth from Calicut in south-west India, studded with pearls and precious stones;elephants,parrots,sandalwood,peacocks, hardwood,incense,tin and cardamom from Siam (modern Thailand)".(3)

Fig 2 A Chinese spider with a human-like face. February 1935.

Mike Hardcastle.

The story of this freak spider is on the following page.

SPIDER WITH A HUMAN FACE

A spider with a human face has been discovered in Chumatien, Honan (China.)

Its body is small, but its head is abnormally large.

The face of the spider is dead white with black eye-brows and a black nose. White lips add to the horrors of its appearance. It feet are dark brown.

A Chinese found the creature in his home and he has sent it to the People's Education Institution (4)

CHINA'S MONSTER

Loch Ness in the Shade

Gigantic Python?

Kwangtung Province in China has now produced the "Loch Ness" monster, for according to reports received here, says a Canton report from Toy-Shan, a hilly district of the province, 42 villagers have been killed recently by a huge serpent which has terrorized the population of that district for some time past. The serpent was finally shot dead by a man more intrepid than the rest and this individual has now received a large reward from the local Magistrate. There are various conjectures afloat locally about the nature of this death-dealing monster, which some believe to have been a giant python of a type occasionally found in remoter districts. (5)

STRANGE ANIMAL AT CANTON ZOO

A unique creature is on exhibition in the Canton Municipal Zoo. It has just been brought here from the mountains of the Taishan district. It has the head of a cat, the body of a horse and weighs more than 100lb, says a Canton report. It is attracting considerable interest. (6)

MAN KILLED BY SHARK AT HONG KONG

From Reg Harris   Hong Kong

Police-sergeant H.W Jackson was the victim yesterday of the first known attack at Hong Kong while he was bathing at Tweed Beach shortly before dusk. He was frightfully mauled and died within a minute of being rescued by Captain A.W Braude, of the Hong Kong Telephone Company. Tweed Beach is a popular swimming area near Stanley internment camp. Sergeant Jackson, who was awaiting repatriation to London. Large sharks have been seen in Hong Kong bathing waters. It is believed that this one must have followed a ship in.(7)

REFERENCES

1 The Straits Times July 15th 1887 p. 2

2 Daily Advertiser (Singapore) May 18th 1893

3 Gavin Menzies 1421 The Year China Discovered The World (2002) pp 33-34

4 Northern Standard (Darwin, Australia) February 5th 1935 p. 8

5 Townsville Daily Bulletin (Queensland, Australia) August 21st 1935

6 The Straits Times January 7th 1936 p. 18

7. The Argus (Melbourne, Victoria, Australia) September 25 1945 p.20

Communist China,the "Gang of Four" and a "strange animal with horns that was not an ox."

Those of you who know me well will be familiar with my many declarations of sanity on my Muirhead`s Mysteries blog on Cryptozoology Online when I introduce obscure off-beat cryptozoology discoveries. As in long lost archive material I mean. This offering I am about to comment upon is somewhat different (but only a bit!) It dates from a Foreign Broadcast Information Service(FBIS) monitoring report which I discovered online in early March 2012;dated January 18th 1978 when I was a mere child of 11.

What interests me in this case is the combination of mystery animal speculation within Chinese Communist intrigue and the custom of Chinese Communists to use the animal kingdom,both natural and bizarre, in their perpetual struggles. The "Quakers" should not be confused with the Christian international movement which began in England in the mid Seventeenth Century. (1) Nor should the "Gang of Four" be confused with the punk band of that name,who formed in 1977 and released their first single "Damaged Goods" in 1978, but:

"Four leading radical figures[who]played a dominant political role during the later years of the Cultural Revolution. Hardliners Jiang Qing (Mao Zedong's fourth wife), Zhang Chunqiao, Yao Wenyuan, and Wang Hongwen appeared likely to seize power. But several weeks after Mao's death in September 1976, they were instead arrested and blamed for the excesses of the Cultural Revolution. Sentences for their "anti-party"" deeds ranged from death (later commuted to life in prison) to 20 years in prison " (2)

The FBIS report said: NATIONAL AFFAIRS. LIBERATION ARMY DAILY DISCUSSES `QUAKERS` AMONG GANG OF FOUR.

21

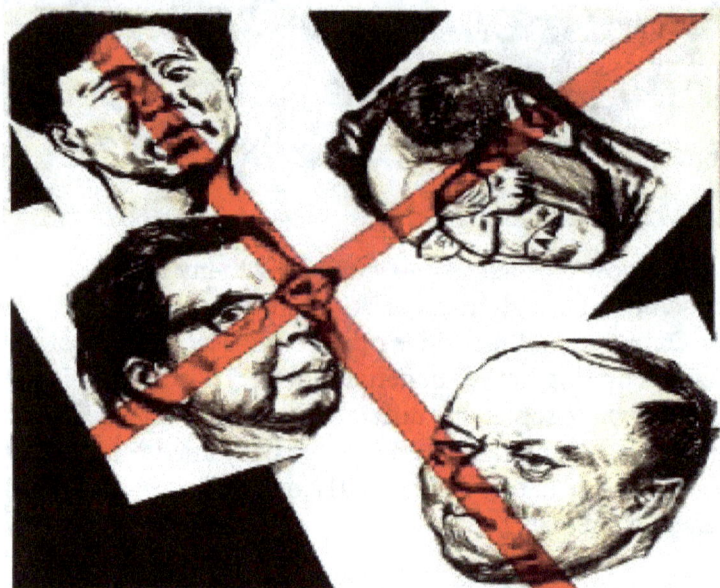

Fig 1 A propaganda poster showing the Gang of Four. The woman (centre left) is Mao's last wife, Jiang Qing. The caption reads- "Decisively Throw Out the Wang-Zhang-Jiang-Yao Anti Party Clique!" Wikipedia Creative Commons.

Peking, 17 Jan - - An old scientist humorously declared : "The `gang of four` bred a strange animal with horns that was not an ox, for the animal has bristles which the ox does not have; however, it was not a porcupine because a porcupine has no horns." This was well put. Among the "gang of four`s." sworn followers, confidants and lackeys, those strange creatures were quite dashing and exciting. Obsessed by their lust for power, they acted as if they were crazy drunk. Their harm was in causing "quakes", so we might as well call them "quakers". As a result of the "gang of four`s" instigation and commands or as a result of their counter revolutionary ultra rightist revisionist line, "political quakes" have continually occurred beneath our feet during the past few years. The sinister military adviser Chang Chun-chiao blared :" We should send those persons with horns on their heads and bristles on their bodies to the central authorities to kick up a row there." (3)

And so it goes on in a similar vein. Deng Xiao-ping is famous for saying, in 1962 " It doesn`t matter if the cat is black or white as long as it catches the mice" in other words it doesn`t matter how China becomes prosperous, through communism or capitalism, as long as it does. So China is no stranger to combining animal mysteries with significant political events.

REFERENCES

1 Wikiredia Quakers http://en.wikipedia.org/wiki/Quakers.

2 BBC website :
http://news.bbc.co.uk/hi/english/static/special_report/1999/09/99/china_50/gang.htm

3 Foreign Broadcast Information Service Daily Reports. Peking NCNA Domestic Service 1978-01-17 Liberation Army Daily Discusses `Quakers` Among Gang of Four.

A Living Mammoth in Mississippi, 1818

LITERATURE,
ARTS, SCIENCES, &c.

THE LIVING MAMMOTH.
FROM THE EMIGRANT

Prairie Du Chien, March 1, 1818.

Sir—The present appears to be an age fruitful of curiosities, on the land as well as in the deep. I take the liberty to send you for publication a copy of a letter to Dr. Mitchell, written by a gentleman of New York, who has been for some time past, a sojourner in the wilds of upper Mississippi. The strange account which he there gives of the re-appearance of the mammoth, and the causes of so great commotions among the wild animals—will afford to the naturalist matter of deep speculation, and excite in the public generally equal curiosity with the recent appearance of the great serpent on the shores of the Atlantic.

I am, Sir, your obedient servant, &c.

The above report was published in The Weekly Recorder (Ohio) April 17th 1818. Dale Drinnon told me in an e-mail that Ivan Sanderson wrote about surviving mastodons in the Eastern US area in the 1700s in Dynasty of Abu, but most people think that source to be untrustworthy. In President Jefferson's time, some hope was held out that some of the the Pleistocene megafauna survived Out West" but that "Out west" kept going further West as the settlers pushed on until the possibility was no longer regarded possible. There was some ambiguous Native folklore that was collected on the theory, though: but nothing as substantial as any report or rumour of wild Elephants in Mississippi or anywhere else in the Eastern US in the 1800s. As far as my research goes to indicate anyway.

On June 3rd 1818 the National Aegis (Massachusetts) carried a letter from a Dr Samuel Mitchell who proclaimed: "Do not be too much surprised at a mention of a quadruped so famed for its size, and which has long since been considered as extinct….Various Indian accounts have lately reached us of its having been seen on the Big Prairie , not far from the head of the Redwood River. "A now defunct web site that commented on controversial author Gavin Menzies proposition that the Chinese sailed around the world after 1421 and took animals from one country to a another[http://www.gavinmenzies.net/assets/images/spacer.gif] stated that the Chinese may have brought elephants to the Mississippi and Missouri areas. Also, " Mr Stanley was captured and taken by elephant over mountains west of Missouri. "

Furthermore: " Hippopotamus from Africa to China (Beijing Museum - `Western Han c. 208 BC`) Rhinoceros. African Bronze rhinoceros sculpted in western Han period. There is a saddle tied around this rhino. There is also a Mo=Tapir. This bronze sculpture unearthed in the Royal tombs of Han Guo of China of the Warring States era is skillfully designed, lively, complex technical achievement according to the experts. There is a Chinese lady sitting on its back carrying a pole attached to a round dish on top. This art is held in Shen Xi province museum. If China did not have tapirs just as it does not have rhinos, Zhou Chinese must have sailed to continents afar to fetch these animals, just like the giraffes delighting the Ming Royal courts."

" There is also a legend popular in Oregon historical circles that a rhino was found on a Pacific beach (USA) , that was being transported back to China but was there due to a shipwreck."

A Dog-Headed Sea Monster off Boneventura, South America, June 1721

I found this story on a Newsbank database of American newspapers-namely The Boston News-Letter, dated Monday June 26 to Thursday June 29th 1721. Bonaventura is probably the main sea port of Colombia.

Paris April 7 [1721]. The Publick here has been entertain'd with the following Strange Account of a Sea Monster, which was seen on the 18th August 1720 in the Gulph of South America call'd Boneventura. That Monster had a Head like a Water Spaniel with an indifferent wide Mouth, broad flat Teeth, fiery Eyes like those of an enrag'd Person, lank Hair, a large flat Nose, Hands,Arms, Shoulders, and all Motions like those of a Man, a brown Skin, full Breasts, like those of a Nurse. In what distinguishes the 2 Sexes resembling a Horse, as near as could be guessed by the Eye he was about 8 Foot high. He was from 10 in the forenoon till 12 within Arms length of a French Ship. The Captain order'd his Men to endeavour to strike him with a harping Iron but he escap'd twice by diving under water, however some time after he came up again, snatch'd a Line out of the Hand of one of the beholders, and moved off Swimming as a Man; then drew near to the Ship again, and raising himself so high above Water that his Knees were seen, so far forgot the Respect due to his Tarpaulian Spectators as to do,what, for Decencies sake, the Author of the Relation has thought fit to leave unexprest and then disappear'd quite. The Duke Regent desiring to be satisfied of the Truth of this Story, sent for the Captain and Mate of the Ship, who assur'd him that the Description afore said is entirely conformable to Truth, whereupon His Royal Highness has ordered a Draught of that Monster, to be deposited among the Historical Records of Paris and with a full Account of all the foremention'd Circumstances. This Monster seems to resemble one which was kill'd in 1717 on the Shelves of Boulogne, by Mr Caron.

A VISION OF THE VIRGIN MARY AND CHILD ON THE ROOF

OF A CARAVAN, ISLE OF WIGHT, EARLY 1980s

Fig 1   This image appeared on the roof of a caravan of some hermits on the Isle of Wight in the early 1980s.

My mother and I knew two hermits who lived in a caravan or portacabin-type structure near Wolverton Manor, the Isle of Wight for a while in the 1980s. They passed on a piece of paper with the above image on it which appeared on the roof of the caravan, although I never saw it myself and I do not know any other circumstances. However, the notelet had the following paragraph on it:

" Just another portrait of Mary? Not quite- This one appeared and stayed for two months on the ceiling of an island hermitage soon after it was established. As it began to fade a need to draw it was felt so as to preserve the image and that the blessing might be shared..."

Fig 2   Google Earth image of Wolverton Manor,Isle of Wight

NATASHA: THE MONKEY THAT WALKED UPRIGHT

ZVI RON

The most famous resident of the Safari Park in Ramat Gan, Israel, is Natasha, a black macaque monkey. News reports quickly spread around the world when she began walking exclusively on her hind legs, with a "ramrod straight" posture like a human. Monkeys generally alternate between walking on all fours and upright movement, but Natasha only walked on two legs.

This unusual behavior began after she nearly died from a stomach ailment. Four monkeys at the zoo were stricken with severe stomach flu. They were treated at the zoo clinic by veterinarian Yigal Horowitz. "I was sure she was going to die," he reported. "She could hardly breathe and her heart was not functioning properly." After intensive treatment, the monkeys pulled through. However, Natasha began walking upright. None of the other monkeys who contracted the stomach flu displayed any unusual post-illness behaviors, and other than walking upright, Natasha acted completely normal. Horowitz said he never saw or heard of anything like this before and speculated that it may be the result of brain damage, or some other residual damage to the nervous system. [1] This story was reported all over the world, accompanied by a striking photograph of Natasha walking upright. A video of Natasha scampering about on her hind legs can be seen on You Tube.[2] The story inspired numerous headlines along the lines of "Missing Link?" and "Monkey Apes Humans", and a certain amount of discussion about evolution [3] and even political commentary. [4] After the initial news report not much was heard about Natasha. The zoo reported that Natasha walked upright for only four days, after which she returned to standard macaque movements. She has since given birth multiple times and lives a normal monkey life at the Safari Park. [5]

What lessons may be learned from this episode? Firstly, that severe illness may cause temporary damage in monkeys causing them to walk upright. This may be a factor in analyzing reports of apelike hominids, they may just be regular simians recovering from an ailment.

29

Although Natasha's case is said by zoo officials to be the only one on record of a monkey walking upright as a result of illness, it is clearly something demonstrated to be possible and may explain some unusual sightings. Secondly, and perhaps more significantly, this case demonstrates the way in which anomalous phenomena are often reported. The initial news report of a walking monkey made a lot of waves and can be found on many news sites as well as sites devoted to the discussion of strange phenomena. However, the follow up report, that this effect lasted for all of four days and everything is now back to normal, was only reported by the Safari Park itself and not picked up by any news agencies. The item appears only in Hebrew and seems not to have been translated or referenced in any forum discussing the case. Recognizing that the upright walking lasted less than a week casts the entire episode in a different light, a temporary unusual side effect experienced while recovering from a severe illness, rather than a true change in the monkey's nature and behavior. Once again we see that when it comes to anomalous phenomena, repeating the splashy initial reports is not at all enough, and a follow up is needed to get a better understanding of the phenomena at hand.

NOTES

1. "Monkey apes humans by walking on two legs", Associated Press, July 22, 2004.
2. http://www.youtube.com/watch?v=8L_SjngmnO8
3. The Israeli religious children's magazine HaChotam pointed out that a monkey walking upright is not evidence of man's evolution from primates http://www.hachotam.org/he-il/kids/articles/science-and-faith/monkey-see-monkey-do/
4. "Israeli Monkey's Lessons in Evolution", National Public Radio, July 23, 2004. http://www.npr.org/templates/story/story.php?storyId=3613530
5. "Natasha the celebrity monkey gives birth again", news release from the Safari Park, Ramat Gan, March 1, 2007. http://www.safari.co.il/arab/article.php?&id=284&prt=1

Fig 1 Natasha, from primates.com

BRITISH BALL LIGHTNING

REPORTS 2009-2011

Sam Hall and TORRO FORUM

The British-based TORRO Forum(TORnado Research Organisation) is an online group devoted to reporting meteorological events from a historical perspective and also as and when interesting events occur, particularly tornadoes and other varieties of severe weather. One phenomenon reported is ball lightning. This essay explores some ball

lightning and GLO (Globular Light Emitting Object)events between 2009 and 2011. Peter van Doorn, TORRO` s ball lightning moderator is responsible for posting the reports on the Forum. The reports have been abbreviated in some places.

Fig 1 A Nineteenth Century depiction of Ball Lightning. Wikipedia Creative Commons

BALL LIGHTNING © RICHARD MUIRHEAD

Luminous, I saw you drift down this ancient corridor,

entering the professor`s study almost gleefully.

Then suddenly you made a sound like a clap of thunder,

leaving an acid smell.

Whatever you are, please don`t come this way again. I won`t be pleased to see you.

During a severe thunderstorm at approx. 02:10 GMT, 26.11.2009 a house at Hastings, East Sussex was struck and the roof slightly damaged: opening found in tiles & lining felt forced out over guttering allowing water into house.

Female resident heard explosion & sound as if large vehicle had entered the lounge-went to investigate & saw a clear globe of light about the size of a football with what looked like a miniature tornado revolving inside.The GLO moved " slowly and erratically" and suddenly vanished without sound

Fig 2 The Great Thunderstorm, Widecombe, 1638,
Wikipedia Creative Commons

The GLO was in appearance like a clear glass ball and not fiery or lightning-like.

This incident was reported to the local authorities who in turn contacted me (rec. 31.12.2009) This is a very important case and I hope to be able to interview the witness in the near future.

Maidstone, Kent-7th March 2010. Location of sighting:Maidstone,Kent

33

Date of sighting: Sunday 7th March 2010. Time:Approx 8.15pm

Witness Name: Jane Davey. Witness Statement: I saw a small white ball in the sky (which definetely wasn't a plane) and as it passed over the house we had a power cut covering our road and beyond. Power was restored after 20 seconds, although the internet and cable tv stayed off all evening. Sidney the cat was spooked for the rest of the evening and wouldn't settle.

Source: www.uk-ufo.co.uk

Comment: If you can provide further information on this or other possible UFO sightings in this area then please leave a comment or send details through our " submit sighting" form.

Comments on " Maidstone, Kent-7th March 2010"

Stephen Saunders March 11th 2010 at 2.40am That's a very interesting (and credible!) report. It sounds like a possible GLO/ball lightning event which could well have caused electricity and telecoms outages.

Here's a link to an interesting paper on the subject:
http://www.torro.org.uk/site/ball_info.php

Danielle March 12th 2010 at 4.52pm That's very interesting..I had a power cut, half way through a programme and I live in maidstone! That is quite cool!!!

Lindsay Hawthorne March 17th 2010 at 5.42pm 7th March 2010 was the same night my mam seen about 10 orange flying objects in the sky red orange/red sightings we live in Blyth Nortumberland.

Ball lightning (?) Colne,Lancs 18th August 2010

Jonathan Webb. 21/8/2010 Not quite sure how to interpret this report but it looks worth following up.

- - - - Original Message - - - -

34

This ` lightening ball` was observed in Barnoldswick, Colne,Lancashire. No thunder or rain showers at the time. Sky was dark and fairly clear.

Barbara Davison.

Dear Barbara

Many thanks for reporting this. Can confirm where you observed this and whether there was any thunder or rain showers around the time.

Best wishes

Jonathan W

Truly amazing! Last night (18th Aug) 9.22pm saw what I can only assume was a lightning ball.

Height was about two terraced houses. Orange/yellow ball which seemed to be inside a black circle (balloon). Had white streaks following behind it like a meteor. Thought it was at first. Travelling horizontally to the ground. Watched for about 15 seconds till it disappeared into the ground behind houses. Was expecting a big explosion but nothing at all.

Can`t believe what I was seeing but that was it.

Barbara Davison

Matthew Harris   22/8/2010

I was in Lancs that week, by the coast and saw numerous Chinese Lanterns.

Peter van Doorn 5/9/2010

Have just read this having returned from Italy late yesterday. Not possible to draw any conclusion other than it was a literal UFO -

may have had an artificial origin-certainly not classifiable as `BL` or a GLO-possibly a small hot air or fire balloon, possibly an unknown natural phenomenon.

Unexplained lights sighting - 6 May 2011

Date - 6 May 2011

Time - Approx. 18:40 BST

Observation location - St Anns Chapel, Cornwall PL18 9HP (looking south)

White lights/object location - estimated to have been over Saltash, Cornwall.

I have just spoken to my parents who live in St Anns Chapel in Cornwall. They were looking towards the south coast at the approaching rain when they observed two white lights/white objects in the sky some distance away. They estimated them to be located over Saltash approx. 12 miles away.

The lights/objects appeared to be approx. 1000 feet off the ground and were circling each other for about 30 seconds. One of the lights/objects then split away from the other and accelerated at a phenomenal speed horizontally in a westerly direction before it was lost from view in the grey cloud. It was estimated to have travelled over 10 miles before being lost from view over the Looe area. The other light had dissappeared when they looked back towards Saltash.

My father who has seen many fighter jets over the years said that the acceleration of the light/objects must have been quite large to have seen them from what he estimates was a distance of approx 12 miles, and would have likely been seen by others.

The sighting was just as the `spanish plume`† rain area was hitting the area.

Regards **John Pask**

Nicola Parsons Subject: Re: Unexplained lights sighting-06 May 2011

John, this is really strange as I have just had a similar experience, just around the time of a huge clap of thunder around 23.00...looking southwest, I only caught a brief glimpse and if it wasn't raining and extremely cloudy I would have thought I had seen a small meteor from a shower, but I know that is highly unlikely...I am wondering if it could be cloud to cloud lightning????

John Pask

Nicola, The white `blobs` my parents saw slowly circled each other for 30 secs before one accelerated off away from the other one. Wouldn't have personally thought cloud to cloud lightning would act in such a manner , but I am no expert-any experts out there who may be able to help?

John

Nicola Parsons

It is very strange, I even had a neighbour comment on it this morning, they too saw a similar thing....I have no idea, it was in the direction of some electricity pylons near me, so I am not sure if they are involved....it would be good to have an expert opinion. :)

Jonathan Webb

Thanks for reporting this, John. It's interesting that it was some time before nightfall which makes some non meteorological explanations less likely. While another cause may come to light, there are some ball lightning incidents on record with similarities.......some involving aircraft. Will be interesting to hear any further accounts.

Best wishes.

Peter van Doorn Thanks for sending in this report John, though this

37

phenomenon clearly exceeds a meteorological solution . These lights
were true UFOs: Unidentified aerial phenomena - I certainly do not
believe that alien spaceships are visiting the earth, however strange
light-emitting objects of an unidentifiable nature, exhibiting remarkable
manoeuvres, have been witnessed in our skies for more than two
thousand years.

These are the salient points:

" The lights/objects appeared to be approx. 1000 feet off the ground and
were circling each other for about 30 seconds. One of the lights/objects
then split away from the other and accelerated to a phenomenal speed
horizontally in a westerly direction before it was lost from view in the
grey cloud. It was estimated to have travelled over 10 miles before
being lost from view over the Looe area. The other light had
disappeared when they looked back towards Saltash."

The lights were "circling each other" - clarification needed here but this
indicates purpose.

They were visible for c.30 seconds - a duration so great it rules out
lightning or any known peripatetic meteorological phenomenon

" One of the lights/objects then split away from the other and
acceleration to a phenomenal speed horizontally in a westerly direction
before it was lost from view in the grey cloud."

This clearly indicates intelligent behaviour: If the objects were not
`secret` aircraft developed , say by the USA , then they UFOs of
unknown origin.

The most `advanced` form of BL, the GLO Navigator acts as if it has a
brain , and certainly carries out sophisticated manoeuvres- I regard
UFOs as `all weather` super-versions of the same phenomenon.

I would class this as an unknown aerial phenomenon[1] related to
BL/GLOs-no direct connection with met conditions obvious, but these
may have provided sufficient ambient energy for the objects to have
become
_____
[1] How about angels? Richard.

visible to human eyes. I can only offer an enigma as explanation for a mystery!

John Pask

Thanks for your comments everyone.

Peter, to clarify,the lights/objects were apparently following each other in a small circular motion.

I have looked on a various websites for other sightings but no luck yet! It occurred over Saltash which is a large town and it was daylight so someone else must also have seen them. I will buy the local paper which is out tomorrow and see if it mentions anything.

Peter   John,

Thanks for the additional info.

Any further details from the local paper would be appreciated, as this is an interesting case worthy of record.

CARL'S COGITATIONS: GREAT SNAKES OF THE WORLD PART ONE

CARL MARSHALL

"As to what is meant by weird - and of course weirdness is by no means confined to horror- I should say that the real criterion is a strong impression strong impression of the suspension of natural laws or the presence of unseen worlds or the forces close to hand." H.P Lovecraft in a letter to Wilfred Blanch Talman August 24th 1926

This article is not to be taken as a definitive coverage of outsized snakes of the world; Rather just a few examples of some giant cryptozoological species as well as some personal speculation that at least one known species may have possible sub-specific variations of even larger proportions.

During my six years as curatorial assistant at Stratford Upon Avon Butterfly Farm I have had the opportunity and privilege to work with many unusual and often deadly species. You see, its not all pretty butterfly's, our latest "nasty" is a Brazilian Wandering Spider; *Phoneutria sp.* (the murderess) and she really is evil.
We also get donated snakes, usually about eight - ten a year; *Colubrids, Boids and Pythons* - of which some of these reptiles are brought in by the RSPCA due to neglect by owners who are ignorant of the care required or lack the sensitivity and commitment. They are also sometimes handed in by the owners themselves who have had to give up their pets often because the snake grew bigger than they were expecting.

The largest snake I have personally worked with was a 15 ft (4.5 metres) reticulated python which despite its large, but certainly not giant size, was very difficult to deal with because of its immense strength and aggressive nature. So now the question we pose - what happens when an owner buys a potential giant like a reticulated python as a "first snake" and it rapidly grows into a hyper aggressive monster? Pet shops do sell gigantic snake species like anacondas and reticulated pythons to

individuals who (at the time of writing) do not need any specific
licensing to buy or keep them, so again,what happens when a snake
exceeds 20ft (6 metres) in length and has the appetite to match? the
owners have four choices open to them a) They can either:

a) Adapt and keep the snake.
b) Sell or relocate it to a responsible carer.
c) Have euthanasia preformed by a qualified veterinarian
c) Illegally release it into the wild.

Unfortunately the latter has happened around the world where there are
many instances of invasive snake species damaging the ecological
balance of the area that they have escaped or been released into.
Currently the Florida Everglades in North America has become home
to the Burmese python which is obviously not its country or even
continent of origin and subsequently they are causing all kinds of
problems to local wildlife and their habitats.

It is believed by established zoology that the largest species of snake in
the world in terms of bulk is the green anaconda from south America,
which we know can reach a massive 28-30ft (approx 9.1 metres) long,
although some acknowledge 37ft (11.2 metres). But does this species,
or any other grow even bigger? I personally believe the green
anaconda (*Eunectes murinus*) does, possibly up to about 50-55ft
(approx 16.7 metres) and also the reticulated python *Python
(Broghammerus) reticulatus* which I feel could, if undisturbed, reach
lengths of up to about 45-50ft (approx 15.2 metres) and the following
article will discuss the implications of these estimations.

South America.

Quoted below is a witness account by adventurer F.W. Up de Graff
and his team in 1923 regarding a colossal green anaconda in its natural
habitat:

"There's a dead alligator over there; let's get out of here."
I turned to look in the direction in which he had pointed. In a moment I saw
his mistake. There lay in the mud and water, covered with flies, butterflies
and insects of all sorts, the most colossal anaconda which ever my wildest
dreams had conjured up.Ten or twelve feet of it lay stretched out on the

Fig 1 The photo of the large yellow snake is of a 13ft albino Burmese python that I am posing with to give an indication of scale. This photograph was taken at Bags o` Reptiles, in Evesham, Worcestershire by the owner and good friend of mine Andy Badland.

Fig 2 The close up photo of a snake`s head is of a Reticulated python

bank in the mud; the rest of it lay in the clear shallow water, one loop of it under our canoe, its body as thick as a man's waist.

I have told the story of its length many times since, but scarcely ever have been believed. It measured fifty feet for certainty, and probably nearer sixty. This I know from the position in which it lay. Our canoe was a twenty-four footer; the snake's head was ten or twelve feet beyond the bow; its tail was a good four feet beyond the stern; the centre of its body was looped up into a huge S, whose length was the length of our dug-out, and whose breadth was a good five feet.

I was in the stern where I couldn't reach the rifles, so I called out to Jack to shoot. He reached out for his weapon, but the noise he made in fumbling for the it alarmed the snake.

With one great swirl of water that nearly wrecked us it vanished. The agility with which it moved was absolutely astounding in view of its great bulk, in striking contrast to the one we skinned. When I thought of how the latter's decapitated body had coiled round my legs and nearly broken them in the last contraction of its dying muscles, I wondered what would have happened to us had that huge beast in its headlong flight taken a turn round the canoe. How utterly helpless the mightiest of men would be in the coils of such a monster!"

Up de Graff. F. W. Head-hunters of the Amazon, 1923.

A snake of 60ft (18.2 metres)? I personally think a snake of this size is possible, yet improbable. There are even reports of anacondas of 150-200ft (approx 60.9 metres) but these are reports from frightened natives and non-scientific observers and should not be accepted at face value without sufficient corroborating evidence. One question we must ask in order to get to the bottom of the question of maximum size in any species of snake in the wild is how long do they live? Although its true that snakes never stop growing this rate of growth does slows with age, so there are definite limitations depending on the age expectancy of individual species. We know that anacondas in captivity can live for 30 years but we really don't know for sure about wild specimens.
Of course resources available, eg. prey animals and the nutritional quality of these will also have a big part to play in determining a definitive maximum size.

Green anaconda:
Snakes are surrounded by legends and mythology. In the bible the book of Genesis says that the serpent is the evil creature that deceived Eve into tasting the forbidden fruit and thus being the reason why man was expelled from the Garden of Eden.

43

In mythology however the snake has not always been seen as a symbol of evil.

Fig 3 The painting of a giant anaconda by Maureen Ashfield

In ancient Canaan it was the symbol of the god Eshmun, the equivalent of the Greek Aesculapius, the god of healing and connected to the underworld and reincarnation due to the snakes ability to cast off its old skin to make way for the new. This author found that In Belize in Central America locals believe the Wowla, which is the local name for the Common Boa Constrictor, is the mother of all snakes simply because it is the largest snake the peoples regularly come across, so therefore must be the mother to smaller snakes even if they belong to an entirely different variety. They also firmly believe that the Wowla is venomous after sunset. When the first Spanish explorers arrived in south America they named the green anaconda el Matatora; the bull killer.

Are we to believe that the early Spanish explorers really witnessed the spectacle of a giant anaconda constricting and consuming an adult bull or was this name purely the work of over-active imaginations? It is widely known that anacondas do catch and consume capybara, wild pigs and even on occasion Jaguars, so why couldn't a giant individual living in some rarely used, slow moving back river consume prey of bovine proportions?

Anacondas are mainly aquatic, but when this species does crawl out onto the land it struggles to pull its massive bulk around. Its circulatory system is also under great stress when crawling about terrestrially due to the snakes emmence weight (this would normally be supported by the specific gravity of the water). The method used by this species to move terrestrially is called Rectilinear locomotion or "the caterpillar crawl", the snake uses two powerful opposing muscles connected to each rib to pull its body along, this form of locomotion usually accompany's lateral undulation, the ribs do not actually move only the muscles beneath the skin. The normal method used by smaller snakes is by actually walking on their ribs; which are again connected to muscles that are attached to the scutes or belly scales, however smaller snakes do also use rectilinear locomotion. When in the water the anaconda moves with incredible agility and grace, with blood circulation also being significantly increased.

For this reason alone I am prepared to believe that the real giant anacondas are likely to spend more of their time in the water than their common counterparts, only really crawling onto land when thermo-regulating, attempting to locate another water source during the dry season or locating terrestrial prey items, so therefore spending much of their time hidden in dark murky swamps, undetected and away from human activity. If this theory is correct then this is a positive outcome because this species is like many animals around the world is declining in overall number and maybe facing extinction. A green anaconda of 30ft+ would most likely be a female as males do not usually attain these sizes.

"The acquisition of energy in the natural world involves a complex interaction between the biophysicle environment in which an animal lives, resources available and there distribution, the social system and how it might constrain access to resources and consequently mating success, and the risk involved in acquiring resources".

Where better to for a snake to achieve colossal sizes and successfully support their weight, than in an aquatic environment? Let us hope they remain there indefinitely.

Giant Anaconda
Sucuriju Gigante:

As mentioned previously, zoology recognises the green anaconda to be the largest known species of snake in the world, and this is likely to be correct but is there another larger ecotype somewhere in south America's vast rainforests and backwaters? some believe there is. What of the Giant Anaconda? The Giant Anaconda or Sucuriju Gigante is reported to be a colossal anaconda type *Boid* which may (or may not) be a new species/ sub-species. It has also been claimed to be a living descendent of the prehistoric *Gigantophis garstini* from the Eocene epoch (54.8 - 33.7 million years ago), However we must consider *Titanoboa cerrejonensis* from the Paleocene epoch (65.5 - 55.8 million years ago) as another suggestion simply because *G. garstini* was from what is today Egypt and Algeria, whereas *T. cerrejonensis* is known from Columbia, this alone makes this more likely even though this species apparently became extinct a considerable time before *G. garstini*. My personal belief however is that it is far more plausible that this enormous anaconda type snake is just that - an undiscovered variety of the green anaconda.

Maybe one day a lucky explorer will get the chance of a lifetime and actually observe and clearly film one of these giants basking in the early sun on the waters edge and whether at 50ft (15.2 metres) or 150ft (45.7 metres), it will be a great discovery and one that hopefully will permanently link the discipline's of natural history and cryptozoology for good.

The Camoodi; a giant horned anaconda is also reported but these could just be very large green anaconda's that have developed extensively wide heads with age and the paired light and dark eye stripes that normally run back at slight angles behind the eyes have moved even closer onto the top of the snakes head as the jaws widen, creating the impression of horns when viewed briefly from above when travelling through water. However the Camoodi is usually reported at lengths of only about 20ft (6.09 metres) not a considerable size for the green anaconda, so maybe its something more interesting after all.

Britain.

Monster in the Thames.

In February 2009 Abraham W. claimed that a photo he had posted on
the Mysterious Britain Website was that of a genuine giant snake
swimming in the Thames. The enormous river snake is claimed to be
real and it does appear that ships in the photograph are having to swerve
in order to avoid contact with it. One on-line reader stated that it
resembled a pipe or mud bank more than a mystery serpent but I think it
looks far more like the work of the Photoshop computer program.
Another very similar photo also released in 2009 reportedly showing
another giant snake in the Amazon in South America which again looks
to be a digitally enhanced hoax.On some web sites this photograph was
claimed to have been taken in Borneo.

There have been many examples of escaped snake species in Britain
existing in the wild , here is a brief list of some examples;

In 1966 on Canterbury Road in Croydon, the RSPCA were informed
that a 10ft (3.0 metres) python of unknown species was on the loose in
a garage building and was surviving on rats. Firemen were called in to
remove the floorboards but no snake was discovered.

An African rock python was discovered in a pillow case in a lane in
Walker, Newcastle in May 2012. A 12 year old boy bravely took the
animal to a local pet shop where it was described as very aggressive.

A Burmese python was seen in the morning in the undergrowth at
Lingswood in Northamptonshire in March 2010.

A 12ft (3.6 metre) long python of unknown species was rescued by the
fire brigade after it got stuck under a shed in a garden in Cornwall in
April 2010.

A woman with a phobia of snakes (*Ophidiophobia*) found two living in her flat in Bournemouth. The first a 4ft (1.2 metre) corn snake (formerly named *Elaphe gattuta* but now *Pantherophis guttatus*) was caught and taken to a local pet shop, the second a similar sized Honduran milk snake (*Lampropeltis triangulum hondurensis*) was killed with a hammer after biting the woman's daughter twice on the hand as she tried to catch it. Both these snakes are believed to have escaped from a local pet shop.

WARNING:
The Honduran milk snake displays very similar red, black, yellow bands to the highly venomous Coral snakes of North America, and actually the very same as some other coral snake species from elsewhere in the world and unless a 100% positive identification can be made, snakes of this appearance should not be handled by the public. There is a rhyme that can help differentiate the harmless milk snakes from highly venomous North American corals:
"Red to Black venom lack, Red to Yellow kills a fellow".

Police have warned parents to keep children and pets indoors after a hungry 7.5ft (2.2 metres) Boa constrictor called Diva escaped from her vivarium and her owners home in Broom crescent Ipswich.
Snakes native to Britain are adapted to living in colder conditions. Any tropical species that ends up out doors in Britain will need to find an area to affectively regulate their body temperature in order to survive (eg. move, shed skin, catch and digest food), which is what some of the previously noted escapees were attempting. I have recently been informed that a large Burmese python is surviving down a rabbit warren near the city of Birmingham feeding on the rabbits, but I have not yet been able to validate this, however we can safely surmise that this is a very unlikely story as the Burmese python would not survive well in these conditions as

big snakes tend to need a more continual heat source to metabolise. So far it appears some smaller individuals are actually surviving here.The Aesculapian rat snake (*Elephe longissima*) has been living well in a small area of Wales since the 1970's with a steady population at Colwyn Bay. Its believed that the British ancestors of these snakes escaped from a local zoo where they used to be imported from Italy. There have been no serious reports of this invasive species outside Wales.

One species put forward as a potential survivor is the Amur ratsnake (*E. schrencki*) as they can digest food and generally do well at temperatures of about 75 f (23.9 c) and hibernates for 3-5 months of the year so could over winter underground in a rabbit warren where it is a fairly constant 55-57 degrees Fahrenheit (13-14 degrees Celsius) and possibly survive and breed here indefinitely due to the thermal inertia of the warren.
So what about large snakes surviving here?

Well this is very unlikely, even though its true that larger snakes would take longer to cool and therefore could remain active longer than smaller species in the British climate it is also true that they would also take longer to re-heat and they would have to find quite a considerable heat source to successfully achieve this, such as a man made heating system. This is why I believe its more likely that big snakes are to be found in and around, or even underneath city's rather than in a natural environment.

As we are all aware, climate change is increasing the overall yearly temperature which will eventually make it easier for these animals to live here. Currently though it is very difficult for most non native snakes to survive , even ones that could potentially tolerate relatively cool climates such as Corn Snakes, as they do not tend to attain the ambient heat available in order to successfully digest food; put simply the digestive process slows drastically as the snake cools and the meal rots within the snakes gut leading to slow and painful death.

TO BE CONTINUED…

More Flying Lizard Reports
From Australia and beyond

There was a report in The Courier-Mail, Brisbane on February 23rd 1934 (via Trove, Australia) as follows: **Maryborough**- "Mr W.R Hetherington,of Torquay,brought to Maryborough a specimen of the flying lizard about four feet in length which was captured near Saltwater Creek. The flying lizard resembled the better known frilled lizard. When aroused it exhibited small fanlike wings behind the head. Its captor had no doubt about its identity, a similar specimen caught previously having been sent to Brisbane for identification."

I find the report below interesting in the Cairns Post of September 13th 1948 (also from Trove.) because , if this is not referring to a fossil, then was it something more substantial,closer in time to 1948?

Moscow (A.P.) The newspaper "Kazakhstan Pravda" reported that the complete skeleton of a pterodactyl - a flying lizard - has been discovered in the Tamir district of Kazakhstan.

This is the third specimen of the ancient ancestor of birds to be found in Russia. The bones are now being studied by scientists.

The Uncertain Fate of Steller's Sea Cow

Richard George

The belated discovery and premature extinction of that placid herbivore, Steller's Sea Cow, *Hydrodamalis gigas*, is one of the saddest stories ever told. So when, as a student (not of zoology, I must add), I read an account of its possible survival in a book by Janet and Colin Bord (1), I was riveted. It still captures my imagination like few other Fortean topics.

The chronology,briefly is as follows (2). Steller's Sea Cow was christened by legendary naturalist Wilhelm Steller, in 1741,on an expedition to Kamchatka in the Russian Far East led by Vitus Bering. Unfortunately for the creature, it could feed more than thirty men for a month,was delicious (it tasted like veal, apparently) and its fat produced flames, and hence heat and light, without smoke or odour, an important consideration in a cold region in the days before electricity.

Officially, the last was killed in 1768.

The Bords begin by citing a piece by Michel Raynal (3). They claim it claims that there were " numerous" sightings in the nineteenth century. But the bulk of their evidence lies in a trinity of later reports.

In the early 1950s, a whale harpooner told Dr S.K. Klumov about a mysterious animal he had seen several times off Kamchatka, near the Commander Islands." Of course,it is not a whale," he told the academic. " We know the whales by their appearance, by their fins and by their blowing..." It was massive,more than 30 feet long,black,and with no dorsal fin

51

Fig 1 Steller's Sea Cow. Wikipedia Creative Commons

and it was always seen at the same time of year, early July.

In 1962, once again in July, but much further north, off Cape Navarin, a group of witnesses, described as seasoned hunters and whalers, glimpsed a group of animals unlike any known cetacean or pinniped. These were slightly smaller, between 18 and 26 feet long, very dark, with a small head clearly separated from the body, an upper lip overlapping the lower one, and a distinctive fringed tail.

This second sighting attracted serious attention in Russia, to the extent that an article by Dr. Klumov was published in Priroda magazine. (4) Strikingly, on the basis of this sighting, Steller's Sea Cow was included five years later in Richard Fitter's book on vanishing [note Present Participle] wild animals (5). " But" Fitter acknowledged, " the Russian zoologist V.G. Geptner believes this to have a misidentification of female narwhals."

Narwhals? According to my trusty Collins Guide (6) , Monodon monoceros, although it does lack a dorsal fin, is mottled grey and white, not black, and females are only up to 13 feet in length. The larger males are , of course, unmistakable by reason of their horns.

I am the first to admit that I lack Dr.Geptner's qualifications as a scientist. But here, as so often in Fortean topics, the explanation from on high is totally implausible. I also cannot accept the claim (made on the Internet) that these were stray Elephant Seals. These, surely,would once again be too small and the wrong colour.

The Bords' final sighting dates from 1976, and here they nail their colours to the mast: " A corpse was found in the summer of 1976 at Anapkinskaya Bay." Alas, they are stretching the facts.

Their source is Arthur C.Clarke, or more accurately a book in his name written by John Fairley and Simon Welfare (7). They quote an article in a Russian periodical by Vladimir Malukovich (8), which describes a sighting made by Ivan Chechulin, projectionist of the Karaginskaya culture and propaganda team, when he was involved in the annual salmon fishing season off Kamchatka. Just after a heavy storm, Chechulin, projectionist of the Karaginskaya culture and propaganda team,when he was involved in the annual salmon fishing season off Kamchatka. Just after a heavy storm, Chechulin had noticed an unknown animal on a tidal belt. On being shown a picture of Hydromalis gigas, he said " Just the same thing. The same tail, the fore flippers and the head..." He was amazed to be told the creature was extinct."

Scientists then discovered that a *piece of bone* exhibited in a local museum appeared to be from a Steller's Sea Cow that had died about ten years previously .

The Bords conclude on an optimistic note: " In the early 1980s Soviet researchers were said to be actively searching for live specimens off Kamchatka, so it is likely to be only a matter of time before Steller's sea-cow [sic] is pronounced ` alive and well` , and not extinct after all. "There is one more alleged sighting, from 1977, from the Gulf of Anadyr, north of Cape Navarin. A fisherman is said to have touched one.

Sadly, in the 35 years since, to the best of my knowledge, the trail has gone cold. There are two alleged sightings on the Cryptomundo website from 2006 and 2010, both in September, but these are immediately suspect as they are from the Pacific Northwest, off Washington state, and appear to refer to smaller animals.

We should not loose hope just yet. Ten years ago, a Channel 4 series on cryptozoology, *In Search of Mythical Monsters*, pointed out that hunting and trapping alone rarely cause extinction, and also raised the seriously possibility that the giant Short-Faced Bear , *Arctodus simus,* officially extinct for more than 10,000 years, could still be roaming the poorly mapped vastness of Kamchatka.

And as I write this article, the Daily Express (October 5, 2011) is reporting on an international scientific expedition to Kemerovo region, in western Siberia, to investigate almasty sightings. Compared to the accepted date of extinction of *Gigantopithecus* and its ilk, less than 250 years for *Hydrodamalis* is brief indeed.

What also strikes me, gazing at an atlas as the dusk falls, is how much more remote and thinly populated Steller's Sea Cow country is even than western Siberia. Kemerovo is a coal-mining area with several large cities. The Gulf of Anadyr, by contrast, is nearly a hundred degrees of longitude further east, near the Bering Strait, with only a few scattered centres of population: Beringovskiy, Anadyr itself, Ugol`nyye Kopi, Uel `Kal, Egvekinot, Nunligram. The Gulf itself is 200 miles wide. If Hydrodamalis gigas survives anywhere, it is probably here.

Sceptics, however - and people, never let us forget, are allowed to be sceptical - will point out that Steller's Sea Cow was very, very large, and that the bigger an animal is, the less likely it is to escape attention. The Bords themselves are aware of this principle, when on the same page of their book they commit to posterity the immortal insight (with respect to an alleged monster chicken) " It is hard to see how a bird with 8- foot footprints could go undetected for long".

And the Bords' sightings date from the Soviet era, when its Far East was a closed region. Today it welcomes an increasing number of adventurous tourists, and is under the mass scrutiny of Google Earth. If Steller's Sea Cow is still out there, you would expect more witnesses, not none at all.

There is a further problem. The Bords, much as I love their paradoxological gazetteers, have proved themselves unreliable in their transmission of evidence. An alleged sighting and a bone do not add up to a whole, fresh carcass. This means we would have to go back to the original Russian articles, which are hard to find even if one reads Russian (and most of us, myself included, do not). You will have noticed how much of this material is quoted at one or more removes: at times it resembles a set of Russian dolls.

And does the word *propaganda* , with reference to Mr. Chechulin, concern you?

Mark Pilkington, in his brilliant book Mirage Men, makes a strong case that many UFO accounts have been fabricated by the U.S. military and intelligence services to distract attention from top secret terrestrial aeronautic projects.

REFERENCES

1 Janet and Colin Bord, *Modern Mysteries of the World* (1989) , p. 300 f

2 David Day, *Vanished Species* (revised edition, 1989), p.215ff

3 Michel Raynal, " Does the Steller's sea cow still survive ?" , *INFO Journal* 51, pp. 15-19

4 S.K. Klumov, " Do large unknown animals still exist on the earth?," *Priroda* no. 8 (1963), pp. 73-5.

5 Richard Fitter, *Vanishing Wild Animals of the World* (1968) , p.49.

6 David MacDonald, Priscilla Barrett, *The Collins Field Guide The Mammals of Britain and Europe* (1993) , p.173

7 John Fairley, Simon Welfare, *Arthur C. Clarke's Chronicles Of The Strange And Mysterious* (1987), p. 104f.

8 Vladimir Malukovich, " Where are you, Steller's sea cow?" *Kamchatsky Komsomolets,* January 1977

NOTES AND QUERIES

Richard George asks: " Does anybody know of Steller`s Sea Cow sightings from the Russian Far East after 1977?" (See his essay on pp 51-56)

The following story has to be the strangest out of Ireland I have ever heard of and that includes mystery animals, leprechauns, etc: Béaloideas is an Irish folklore journal.

THE TINY FOOTBALLERS

Some years ago there was a man living near Ballyroan who used to work for a farmer in Colt. One evening in summer he was returning home from work. He used to cross the fields as a short-cut, and after he had passed through the bog and entered Ballyroan he saw a number of tiny men kicking a football.

He watched them for quite a long time, and in the end one of them came over and addressed him thus: " Kick the ball, Jimmie!" and at this he made several attempts to kick it , but no matter how near it appeared to him he was unable to hit it. The game continued for quite a long time, and in the end his turn came, and he struck the ball. When he did so, he fell in a dazed condition on the ground, and when he recovered the little men had disappeared. Béaloideas vol 9 1939. (1)

Does anyone have similar stories of tiny footballers?

I found the following on the British Library 19th Century On-line
newspaper archive.

LUMINOUS WORMS

Luminous earthworms (says the " Daily Chronicle") have recently been
seen in Richmond and other parts of the Thames Valley. But it is pointed
out that these phosphorescent annelids are not uncommon, having been
described by Grimm as early as the year 1670. Many marine worms are
also luminous. Many marine worms also are luminous. Mr Hilderic
Friend thinks an occasional phosphorescent worm may be useful to the
race in so far that it indicates to them the whereabouts of the others with
which it lives in union, for though earthworms do not possess eyes,
Darwin clearly proved that they are sensible to light. (2)

Continuing the theme of worms, the following was in The Hong Kong
Telegraph of July 6th 1907:

TRAIN STOPPED BY WORMS

STRANGE PHENOMENON ON THE SIBERIAN RAILWAY

The *Dalyokaya Okraina* (translated in the *Japan Advertiser*) reports that
No 3 post train, before reaching Pogranichnaya station recently, began to
proceed more and more slowly until suddenly it stopped-entirely. The
passengers jumped out from the coaches and beheld a strange scene. The
two locomotives were puffing and hissing, the wheels turned but the train
did not stir from the spot. On examination it was seen that the line was
covered with some kind of green moving mass, which turned out to be
worms, apparently a species of woodworms. They thickly covered the
entire road and thus the locomotives were stopped. The poor passengers
had to walk for a distance of five vents (?) ,which the train covered at a
snail`s pace, the journey taking three hours, while passengers and rail
servants helped to clear the rail of the worms. It would be interesting to ,
says our contemporary, how such a large migration of worms can be
explained, and if a similar phenomena has been observed at any other
time. In Manchuria the older residents might be able to answer this
question. It may be added that in America have been held up in much the
same fashion by migrations of caterpillars (3)

On the I Hate Butterflies Online Community: " Girl swallows butterfly, dies. 08/12/02 15:39 - (SA)
Uitenhage - A nine year old girl has died in Uitenhage in the Eastern Cape after accidentally swallowing a butterfly SABC radio news reported on Sunday. Megan Baartman's uncle, Johnny Baartman says the girl was chasing butterflies when one flew into her mouth. She began choking shortly afterwards. Her mother took her to a doctor, but she later died. It is suspected that her breathing passage was blocked. Doctors say a powdery substance on butterflies may cause internal organs to swell. "
"And there's also the fact that certain species of butterflies and moths can cause blindness. If you rub the scales off the wings and you accidentally rub your eyes, you can blind yourself. It's somewhere in the forums." (4)

Also: **Butterfly cemetery**

by privolavacnavytah » Thu Jan 05, 2012 10:55 pm
I never knew, why I'm so afraid of these flying creatures. I can remember as a small child, I was even chasing them in a fields and giving them freedom again. But now I'm scared so much I can't even go out to the nature in the summer, cause they are fluttering all over. I was thinking what exactly caused this fear and I can recall on one story, from my later childhood.
I was around 8 years old, playing hide and seek with my cousin in our grandma's farm-house.
When I was about to hide, I found really great place where my granny kept her farm equipment. So I jumped in there very quickly and squatted down not to be seen.
Then I opened my eyes and in that moment, I saw hundreds - thousands of dead butterflies and wings. All over the ground around me!
All of them were one sort. The peacock butterfly with deadly blue eyes on it's wings.
As I jumped out scared it made an air swirl and dead wings were flying all over as well on MYSELF!!!
I know from that time I never entered places like that one was. With time, I forgot experience totally..
But now, I think this was the cause of my crazy fear, because the most

I know from that time I never entered places like that one was. With time, I forgot experience totally. But now, I think this was the cause of my crazy fear, because the most awful and scariest butterfly for me is the peacock butterfly :(

Privolavacnavytah (5)

Re: Butterfly cemetery

by KixyBoo » Fri Feb 17, 2012 1:16 pm
That's awful :(i think sometimes the dead ones are as scary as ones that are still alive. (6)

Fig 1   Peacock butterfly . By Korall. Wikipedia Commons. Stockholm, Sweden.

A BRITISH TREE FROG IN WARWICKSHIRE IN 1883

The following report appeared in The Midland Naturalist vol 6 1883

TREE FROG IN ENGLAND - I think it may interest some of your readers to know that my little brothers, when out for a walk near Hampton-on-the-Hill last Saturday, caught and brought home alive a small tree frog, which seems to answer to the description of the green one (Hyla viridis) which is common on the Continent, but which I have never heard of as an inhabitant of Great Britain. The creature was about 1 ½ inch long , the body of a beautiful bright pea-green colour above and white beneath; a dirty yellowish line ran down the side from head to tail. Its legs were green above, and of a dark reddish colour below; each of the toes was furnished near the tip with a flat round sucker similar to those on the foot of a fly.

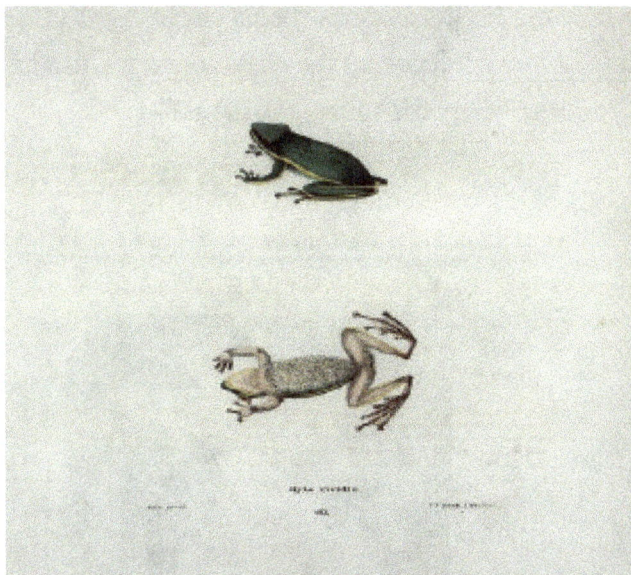

Fig 1

Green Tree Frog John E.Holbrook. N.American herpetology.

The throat was very capacious and hung down like a pouch. When placed in the conservatory it exhibited great activity, climbing and leaping from plant to plant almost like a small squirrel. Fearing it might escape, and until I could ascertain something as to its habits and food, I placed the frog in a large box with some plant, where it remained until

Monday, but then died, whether from want of food or water I do not know. I have now had it preserved in spirits, and shall be pleased to show it to anyone who may feel interested in the capture. All the books I have seen on the zoology of England say that no species of tree frog is known here, but if so, the finding of this one is difficult to account for, especially as it was caught quite out in the country, at a long distance from any place from which it is at all likely to have escaped. - LLOYD CHADWICK, 27, High Street, Warwick, August 23, 1883 (7)

Darren Naish in his Tetrapod Zoology Book One (CFZ Press 2010) comments: " ...It`s also now being suggested that the European tree frog *Hyla arborea* colonies of the New Forest are also natives." [1]

[1] D.Naish Tetrapod Zoology (2010) p.144

Fig 2 A WHITE SWALLOW

FRIDAY, SOUTHERN DAILY ECHO. JULY 16, 1932.

A PURE WHITE SWALLOW

The text under this photo of July 15th 1932 in the Southern Daily Echo reads: " A pure white swallow with pink eyes and pink feet was found at Baskett`s Farm,Gurnard, by Mr.Butchers, the tenant of the farm. The birds had built their nest in one of the pigstyes on the farm, on the ledge of the wall inside the stye, about 6 feet from the ground. The nest contained four young ones, three of which were ordinary swallows, while the fourth one was pure white. They had flown before the photograph could be taken, and had enjoyed some 24 hours freedom. It was hoped they would return to the nest at night to sleep, which luckily they did, and after being caged for a short period till morning while a photograph was taken, they were all again liberated. A white swallow has been seen in the same neighbourhood for three successive years, and the inhabitants are of opinion that this interesting albino is from the same pair of birds. (8)

The Examiner (Launceston, Tasmania) May 30th 1914 reported on a poisonous bird from Papua New Guinea:

THE BIRD OF DEATH

Only one specimen of venomous bird is known to the student of ornithological oddities -the Rpir (?) N`Doob or "Bird of Death", a feathered paradox of New Guinea. Persons bitten by by the creature are seized by maddening pains, which rapidly extend to every part of the body . Loss of sight, convulsions, and lockjaw are symptoms which follow in rapid succession. (9)

The following paragraph is from the Natural History Journal 1888.

FOX HYBRID ?

A STRANGE CREATURE.- Whilst driving home from Chelmsford this morning I saw running across a field a strange animal of a sort of grey colour.It seemed to be something between a little fox and a hare. It did not seem to be the right colour for a fox, yet it had a longish tail tipped with white; as far as we could see too, it had long, brown ears tipped with white. My father, who was with me thought it was a cat; but then a cat would not have such ears, I argued. Then someone else suggested it must be a hare; but the long tail put an end to the suggestion, and we none of us thought it was of the right colour for a fox. We watched it till it was out of sight, but could not make out what it was. When we first saw it it was in a field, running cross-ways, jumping over the borks; [Ridges in a ploughed field]but afterwards, it went through the hedge into a meadow, where it ran into a hollow out of sight. I think it must have been *a young vixen*. Apr 18th, 1886.-*Eva Christy's* Diary. (10)

AN ELECTRIC CAT

From the Manchester City News in the early 1890s:

" There is a cat in a cottage on the Carnice Mount, near Monte Carlo, which seems destined for a show. That electric sparks are evolved from a cat`s skin rubbed in dry air is of course a very familiar observation. But the Monte Carlo cat, according to *Electricity*, is on very dry, dark nights quite a spectacle. Every movement of the body sends off hundreds of minute blueish sparks, something like those discharged by ill (?) adjusted brushes though not so pronounced in colour. They make a noise , on a small scale resembling the crackling of burning furze. Stroking the fur increases the sparkling, and ruffling is the wrong way - which annoys the animal, while the other in no way affects it - produces a miniature pyrotechnic display quite remarkable." (11)

Finally a scare about a supposed gorilla from New Zealand in 1871:

THE SUPPOSED GORILLA

The " Illawarra Mercury" writes:- "Since the report of the strange animal seen by Mr George Osborne on the Avondale Ranges, and which he supposes was a gorilla, has appeared, speculations regarding the existence and species of that `natural curiosity` have been rife in this district. Several parties have been exploring the bush and gullies in the supposed whereabouts of his gorillaship during the past fortnight, but without success. On Wednesday last a party of between 20 and 30 gentlemen assembled in the vicinity indicated, with dogs,

64

ropes and firearms, but after considerable wanderings had been made by some of the party, and much patient waiting on the part of the others, night closed on the expedition without their having obtained either scent or sight of the gorilla. Strange steps and marks were noticed up the side of a fig tree in one of the gullies explored, and as the peculiar formation of the tree providing a hiding place near the upper part of the trunk, it was cut down by several willing hands, but on the tree being laid low an 'old man opossum' made his appearance instead of the gorilla. It is hoped the animal may be captured without delay: and as there is some talk of an expedition being started in Sydney for the purpose, the young men in this district should bestir themselves, and not allow others to carry away the praise and profit as well as the prize involved in the strange affair. A person who has resided on the Bulli Mountain for several years positively asserts that an animal similar to that seen by Osborne, but considerably larger, has been seen in that locality more than once, and by different persons, and that no dogs can be found to face it." (12)

BLACKBIRD FEEDING CARP IN OXFORD

The following extraordinary story appeared in The Country Man magazine for Aug-Sept 1997 Wildlife and Tame column. Editor's remarks in italics.

Blackbird feeds the fish Mr H.M. Youd of Wheatley, Oxon, writes: Some years ago, I was working at Christchurch College, Oxford. At the centre of Tom Quad is a large pool, famous for its fountain. The pool contained a number of big carp that rose habitually to the surface for food. One day, to our utter amazement, a female blackbird perched on the edge of the pool with a beak full of insects and began to feed the fish, who gratefully accepted. For about three weeks, this spectacle attracted students, staff and tourists alike. Given that it was May, I presume the industrious bird must have mistaken the gaping mouths of the fish for a hungry brood.

Sadly, I was informed that last year killed all the fish in the pool and the college was uncertain whether or not to re-stock. So it is not clear when, or if, this remarkable event will ever be repeated. [*This is a fascinating case, though not unique; I have seen photographs of a couple of other celebrated examples involving different bird species. As Mr Youd suggests, the blackbird, which probably had fledglings of its own in the vicinity, was almost certainly duped into parting with its food by the carps' gape (which is quite birdlike) If the gape were yellow or orange, it would have presented a particularly strong 'feed me' stimulus to the parent bird.*] (13)

N.B " Christchurch College is now known as Christ Church according to Rob Wilkes, a friend who works at the Bodleian Library.

PYGMY ELEPHANT IN SOUTHERN VIETNAM IN 1794?

There is a fascinating list of the tribute given to the Chinese Emperor in 1794 by the British envoy,Lord Macartney, in the The Daily Advertiser (New York) October 9th 1794,which includes, under the words 'Cochin China' (Southern Vietnam) 'A very small elephant, 14 inches high.'Now if this was just an ordinary juvenile elephant of any species known then, there would be no significance, but the point is this is a list of exotic and special objects, so there must have been something rather special about this particular elephant. (14)

CRYPTIDS OF 1904

The Times of Swazieland November 5th 1904 reported in a story titled **Animals Uncaught**, a round-up of the world's mystery animals of 108 years ago. This gives an interesting insight into which animals were thought to be strange then. They included.

1. A new species of jaguar…It's a big black fellow, and tremendously fierce. Nobody has ever taken one alive.

2. In Burmah somewhere is a rhinoceros that has a black hide and big tufted ears. The hide has been seen by white men lots of times, but they haven't ever seen a living animal.

66

3. Up in the Himalayas a man has been looking for- what do you suppose? A unicorn. He may be crazy - He may be right. He says that he has heard so many tales from the native hunters up there of the existence of an one-horned antelope horse that he is bound to try and get one...

4. Down in New Zealand, comparatively small as the land is, there are many animal and bird mysteries still. They say there is a brand new - that is, new to the world - type of animal on the order of the duck bill down there yet. Darwin always thought that some day a veritable lizard-bird (not a flying lizard, but a true missing link between the birds and the reptiles) might be found there.

5. " One explorer found mysterious footprints in the snow of the high mountains of New Zealand, but never came up with the boast that had made it. But they were such strange footprints that other scientists agreed with him that the thing that made them was quite unknown to the world, and must be a wonderful thing." (15)

Fig 3   Duck Billed Platypus - once a native of

New Zealand? John Gould 1863 Wikipedia Commons

REFERENCES

1   Béaloideas vol. 9 1939

2   Nottinghamshire Guardian April 22nd 1893

3   Hong Kong Telegraph July 6th 1907

4   http://www.ihatebutterflies.com/forums/viewforum.php?f=10

5   Ibid

6   Ibid

7   The Midland Naturalist vol 6 1883

8   Southern Daily Echo July 15th 1932

9   The Launceston Examiner May 30th 1914

10 Natural History Journal 1888

11 Manchester City News November 26th 1892

12 North Otago Times June 16th 1871

13   The Country Man August-September 1997

14 The Daily Advertiser (New York) October 9th 1794

15 The Times of Swazieland November 5th 1904

This is a drawing of an unknown species of crab, drawn for Matt Salusbury by Indian naturalist Sali Palode. Matt believes this gives him permission to allow me to publish it here. It was seen in an inland forest in Kerala state, India. There is also a photo of it on Sali's web site www.salipalode.com where it is shown to be violet in colour. In an article in the Saturday Telegraph magazine of February 11th 2012 `Paradise in Peril` concerning Madagascar,Richard Grant states: " We saw crabs scuttling sideways in a forest many miles from the ocean" (1)

REFERENCE

1. Paradise in Peril R Grant ,Telegraph Magazine February 11th 2012

DEVO JOCKO HOMO

They tell us that

We lost our tails

Evolving up

From little snails

I say it's all

Just wind in sails

Are we not men?

We are devo!

We're pinheads now

We are not whole

We're pinheads all

Jocko Homo

Are we not men?

D-e-v-o

God made man but he used the monkey to do it

Apes in the plan

But who can prove it?

I can walk like an ape

Talk like an ape

I can do what a monkey can do

God made man

But a monkey supplied the glue

BOOK REVIEWS

In the Wake of Bernard Heuvelmans - An Introduction to the History and Future of Sea Serpent Classification. Michael A Woodley Bideford England: CFZ Press 2008 ISBN 978-1-905723-20-1

This is an academic and thoroughly readable manuscript Woodley`s expression) examining sea serpent classification from the earliest days of "the Rafinesque model" of 1819 up to Bruce Champagne in 2007 who " essentially retains the categories developed by Heuvelmans as `archetypes` , but in addition…employs a multifactor, intensely data-driven methodology to build his own categories or `types."After this Introduction comes the Comparative Methodologies section which looks at different ways of establishing the identity of a cryptid depending on the data available, including Woodley`s own "plausibility method" which is hypothesis driven when faced with cryptids about which little is known, such as the Mongolian death worm, (and the Namibian flying snake of course).

A Conclusion summarises these methods and the author states: " The purpose of this manuscript is to use a combination of contemporary zoological knowledge and aspects of the plausibility method to re-evaluate Heuvelmans` final eight cryptid identities , so as to shed light on possible new and alternative identity theories - something that is in many respects long overdue." Following on from this, there are sections covering Heuvelmans` pinnipeds, Heuvelmans` archaeocetes, The super- otter and the many- humped sea serpent: close cousins? Then Marine `saurians`: genuine archaeocetes ? , Super-eels: a many faceted enigma, Giant invertebrates: The most plausible category. I was particularly interested in the con-rit, a.k.a the many- finned sea serpent, as there

71

was a case in 1883 at Hongay Beach, Vietnam and the oarfish,which Woodley places in the super-eel category. Oarfish have turned up in Hong Kong from time to time.

Discovering Natural Israel Michal Strutin New York: Jonathan David Publishers 2001 ISBN 0-8246-0413-X

This must be one of the best books available on the fauna and flora of Israel and at the same time one of the best natural history books generally speaking, of this new century, even though it was published 11 years ago. It is well researched,brings in Abraham, archaeology,Biblical references, historical events, Josephus, natural historians and covers in great detail, (though not in an overly academic way) the natural history from south to north, starting with an Introduction which states " Israel is full of natural wonder. Yet,despite the fact that vegetation clothes the hillsides and fringes the rivers, despite the fact that birds by the millions travel through the land and more than seventy mammal and eighty reptile species live upon it, untwining natural history from human history is difficult if not impossible." The main eleven chapters cover The Southern Negev to Western Galilee and the Coast. There are numerous colour photos and maps and also boxes on some pages which describe in greater detail some animals e.g the camel (p.21), the ibex (p.95) cleric- naturalists (p. 109) etc. There is no cryptozoology in the book, though I was interested to read about the survival of the crocodile to c. 1908 and leopard. The Resources section (pp.301-308) includes nature reserves,parks, Society for Protection of Nature in Israel Field Schools, related sites. Then web sites , bibliography.(pp. 309-311) species names (pp. 313-322 and a comprehensive index (pp. 323-340).

BioFortean Notes 2 Chad Arment, ed. Pennsylvania: Coachwhip Publications 2011 ISBN 1-61646-109-8 This is a worthy successor to volume 1 and contains the following- The Broad River Sea Serpent, John Hairr Probing The 1896 St. Augustine Carcass, Nelson Jecas and Renee Fratpietro, Irish Snakes , Wild Cats and Other Mystery Animals Richard Muirhead , Sonoran Sasquatch ? , Alton Higgins, The Historical Bigfoot: A Supplement, Chad Arment. Most of these are highly reliant on newspaper reports, but that is no bad thing. Jecas and Fratpietro`s account is somewhat technical when it comes to the tissue  72 sampling accounts. Arment`s   supplement is highly detailed.

Letters to Flying Snake

My cousin Daniel Kenning wrote this letter to me in response to my interest in cryptozoology.

Dear Cous,

I was thinking slightly differently-about different levels of existence inside human consciousness. For example, cats are in my consciousness and I believe they exist because I've seen them and touched them.I believe Koalas exist because I've seen photos and movies and read stuff about them,but I have no personal evidence. I have no personal evidence of the Loch Ness monster, but I've heard stories both that it does and it doesn't exist. I've never seen a were wolf,and I've only heard what I believe are made-up stories about them. I've not seen a flying reindeer,and I'm comfortable in my belief that this is a made-up animal. So there are lots of shades of consciousness or belief about an animal between " exists" and " doesn't exist" , as I'm sure you and your colleagues debate that often.

I was thinking about what makes humans define the "existence" of things, and there is an element of us projecting ourselves on our surroundings to make things exist-I don't mean humans imagining things like Puss-in-Boots or the Frog Prince,Shrek or Dumbo (which we know are made up,and I expect those don't count in cryptozoological terms.)

Beyond existences that can be easily proved by science,like cows-although I'm no scientist so I don't know how you'd prove the existence of cows empirically- my question is " what makes a Loch Ness Monster or Yeti come to sit there within a human consciousness?"

73

I don't mean "can they be proven to exist?", or "is there material proof?" but even before there's any proof, if we take "exist" to mean "to be within human consciousness", then is there a range of reasons why humans are conscious of animals ?

For example, the Loch Ness Monster may owe its own particular level "existence" in our consciousness to a few sightings, but dragons may "exist" because ancient humans needed an explanation for dinosaur bones. Another animal may exist because of innate fear (maybe of former predators of humans like wolves). Others may exist because of human emotions - perhaps ghosts due to reluctance to let loved ones go when they die?

I was just enjoying some rambling thoughts as a diversion from work and stress, I hope you don't mind. Actually the same thinking could be directed towards the (what I think is called) anthropomorphism in religion, to explain why the abstract and un-understandable entity that we call God is drawn like a human, because we humans can't actually cope with concepts like "God is good but we can't fathom appearance", so we decide that He's a he, and looks like a nice friendly father figure because we can cope with that concept."

In mid January 2012 I sent a copies of Flying Snake 1 and 2 to Sir David Attenborough and I asked him about the likelihood of any unknown species of animal in Madagascar. He gracefully replied in a hand written letter and part of it went as follows:

Dear Richard Muirhead

"In reply to your letter - Madagascar is still under-researched so there are doubtless many more species still to be discovered..."

So I find that very encouraging.

In late March I wrote another letter to Sir David querying why some crabs could be found so far inland. He replied (in a typed letter this time)

Dear Richard Muirhead,

"Thank you for your letter.

There is nothing strange about finding crabs in the Madagascar forest-or indeed in Kerala. Land crabs are common creatures throughout the tropics and belong to several different genera. The biggest is the coconut crab which is a pest in coconut plantations - and quite good eating. There are sixteen different species of them on Christmas Island, one of which makes mass migrations to the sea to lay their eggs, a spectacle that has been shown on television."

Also in March was the rather surreal "non-revelation" of Madagascar`s mystery animals.

A colleague at work in Oxfam gave me a copy of the Saturday

Telegraph Magazine for February 11th 2012 with a long essay in it on the fauna of Madagascar. I somehow came up with the thought that the author, Richard Grant was the same person as the famous actor Richard E. Grant. So I tracked down Richard E. Grant`s agent`s address on the Web naively believing I had the right Richard Grant. I had not! About one week later I received a recorded message my phone, from Richard E.Grant, politely stating the fact that he knew nothing about Madagascar`s cryptid fauna as he never wrote the article in the first place!

Now I have done some daft things in my time but this has to be amongst the daftest! Say I`d presumed a " B.Obama" was you know who and I`d got a message on my phone from Barack Obama President of the U.S.A. saying there were no living mammoths in Colorado? Way-hey!

On October 10th 2011 Richard George wrote to me as follows:

" The cryptozoology music connection is interesting. You probably know that a band called the Flying Lizards had a hit single in 1979 with a quirky version of the old standard Money. I owned a copy when I was thirteen."

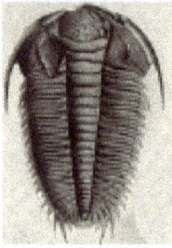

The Steampunk Naturalist

Mad Science and Natural History, Victorian Style

WWW.SteampunkNaturalist.com

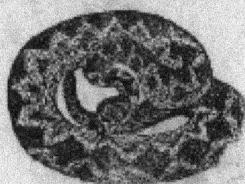

The Rattle-Snake.

The Sea-Orb.

The Green Lizard.

The Torpedo.

Man and Ass.

Hares.

Robinson Crusoe shooting at Muley.

Black Dwarf and the ladies.

Published by W. DAVISON, Bondgate Street, Alnwick.—No. 3.

Print byThomas Bewick, published in Alnwick, c. 1820-1840

Flying Snake

A Journal of
Cryptozoology, Folklore and Forteana

Volume 2   Issue 1   November 2012   £3.99

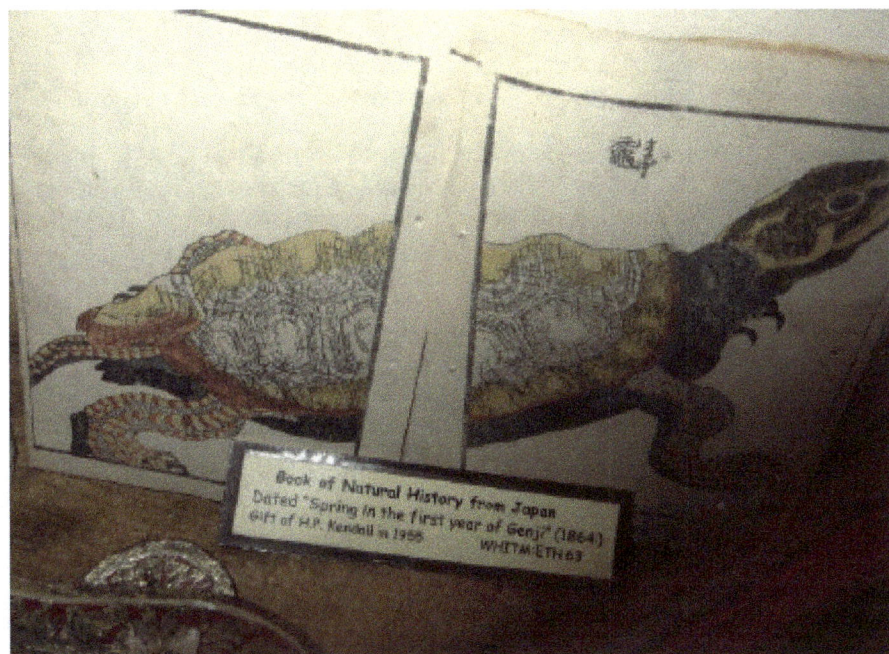

Book of Natural History from Japan
Dated "Spring in the first year of Genji" (1864)
Gift of H.P. Kendall in 1956
WHITM:ETH:63

ABOUT FLYING SNAKE

Flying Snake is available from:
Richard Muirhead
Flying Snake Press,
112 High St,
Macclesfield,
Cheshire,
SK11 7QQ
UK
http://homepage.ntlworld.com/richmuirhead/cryptozoology/

Tel: 01625 869048
Andy Scott Oxfordshire Correspondent
 Mike Hardcastle,Sub-Editor, Australia .   Zvi Ron Israel correspondent.
www.steampunknaturalist.com   Carl Marshall   Zoological Consultant.

Please feel free to contact me if you want to reproduce anything I have written. If you want to reproduce other authors' works, I will try and contact them on your behalf and get back to you. The opinions of authors other than myself do not necessarily reflect my own.

Back issue available on request.

PAYMENT

Subscriptions: £ 3.99 per issue, £ 12 per annum. PDF via e-mail. £3

Payment for however many issues you (and your friendly neighbourhood, reading, flying, snake) would like to purchase can be made by means of PayPal on my web site (See url above).

Checks and postal orders from within the United Kingdom should be made out to Richard Muirhead, and **NOT** Flying Snake.

Checks will not be accepted from outside of the U.K at this point in time.

CREDITS

The image on the cover is from the collection of Dr Devo. See also story on pages 61-62. Those articles which have no author given, are mainly extracts from newspapers, blogs, etc. Thanks to Mailbox,Macclesfield, for printing.

CONTENTS

DR DEVOS DIARY

"For I pray God for the introduction of new creatures into this island. For I pray God for the ostriches of Salisbury Plain, the beavers of the Medway and silver fish of Thames." Christopher Smart , `Rejoice in the Lamb`. Poet,naturalist,lunatic (1722-1771)

It`s that time again when you eagerly flick open your new edition of Flying Snake and I`m glad to say I have a selection of treats for you. Those sharp eyed amongst you will have noticed the mistake in F.S. no. 3, the apparition of the Virgin Mary was on the ceiling,not roof of the caravan. I have a U.F.O. story from near Macclesfield which I was surprised to find has its fair share of U.F.O sightings. I have found or indulged in a number of Fortean stories this time, including a brief study of Chinese coins found far beyond China itself, which just goes to show how advanced the Chinese civilisation is. Whilst we were running around in bear skins the Chinese were posting songs on You Tube! Er,well, slight exaggeration but I hope you get my point ! (Nothing as exciting as S.Korea`s Gangnam Style tho`.)

Flying Snake is not yet available on Kindle as I mentioned in Dr Devo`s Diary in F.S 3 because I didn`t get my act together and frankly the whole process is much harder than I thought. However I do have a medium term plan to do a Wikipedia entry on the Flying Snake of Namibia, my talk on which seemed to go down well at Weird Weekend 2012 in Devon. As promised I bring you Part 2 of Carl Marshall`s essay on Giant Snakes but I`ve held over the story of Ota Benga, a Congolese "pygmy" held in Bronx Zoo in the United States as it was claimed at the time (1906) that he was a "missing link." I also intend to publish a study by veteran Fortean Bob Skinner and myself on the entombed bat phenomenon which has uncovered some interesting data.

My late father mentioned the Mary Tofts rabbits case to me several times and so I`m happy to include it here and as you all no doubt agree with me that the Apocalypse is fast approaching and the only survivors will be cats and humans, I also include,so you can prepare yourselves, a feature on telling the time from cat`s eyes. Miaow! I mean, Ciao!

A Relation of a terrible Monster taken

by a Fisherman neere *Wollage*, *July the* 15.1642. and
is now to be seen in *Kings street, Westminster*.

The shape whereof is like a Toad, and

may be called a Toad-fish; But that which makes
it a Monster, is, that it hath hands with fingers
like a man, and is chested like a man.

Being neere five foot long, and three foot over, the
thicknesse of an ordinary man.

The following Discourse will describe him more particularly.

Whereunto is added,

A Relation of a bloudy Encounter be-

twixt the Lord *Faulconbridge* and Sir *John Hotham*,
wherein the Duke of *Richmond* is hurt, and
the Lord *Faulconbridge* taken prisoner.

With some other Miscelanies of memory both by sea and land,
with some Forreigne Occurrences.

L O N D O N,
Printed for *Nath: Butter*. 1642.

A Relation of a terrible Monster...called a Toad-fish, etc

1642

God sheweth his wonders in the deep, (saith the royall Prophet) but those wonders are never without wonder, when once they leave their wonted stations, and come (upon what message God knows) to visit us in an unknowne world: their shapes being as miraculous to us, as our element unnaturall unto them. But to the purpose. *Friday* morning, *July* 15 between 4 and 5 of the clock in the morning, a little above *Wollage* , one Thomas West, casting his net upon the comming in of the tide, for Salmon;upon the drawing in of the net, (whose weight and difficulty in dragging portended to him good tidings) on a sudden he found a strange alteration: he sees in the net a Fiend, not a Fish; at the least a Monster, not an ordinary creature. Had not his companion had a better resolution, he would have been rid of his net, then troubled with his guest, so deeply was he struck with the odious shape of it. I now proceed to its shape and dimensions. It is by the vote of divers Gentlemen of great quality that went to see it, such a monstrous creature as scarce can be beleeved ever to have been seen: this morning brought alive into Glove-Alley in Kings street. It is called Toad-fish, and with good reason; for the head and eyes, when it lyes upon its belly, doe perfectly resemble a Toad. But here lyes the wonder, turne him up,or but a little raise his head, and you shall behold the perfect breast and chest of a man: nay you may evidently tell as many ribs, both short and long,as are in a man, and of the same joynture and feature; and two as perfect hands as any man whatsoever. By which it is evident that he swims upright, beating the water with his hands,as we all know how the Toad marcheth with his snowt upwards. His mouth very broad, with three ranks of sharp teeth; whereby it is probable that it is a devouring, ravenous, and prey-booting fish; yet is its mouth the very embleme of a Toad, as likewise its eyes. A Butcher's

6

wife coming in hastily to see it, and hearing at the first that there was a strange fish to be seene, and being upon it in the stable where it lay before she was aware, thrusting in among others, started from it with a shreek,crying, *Oh the devill in the shape of a great fish,* swounded, and was faine to be carryed out. The dimensions of the fish are these:

He is in length well-nigh five foot, in breadth a yard over, having on each side two huge fins, in likenesse much like a Thornback, his taile a foot in length, as it were all of a whale bone.

Now the comming up of this monster into the fresh river, and so nigh the shore, is more than remarkable , (never any of this strange kinde ever having beene seen by any age before:) For *Plinie*, the Naturalist, although he confesseth that there is no creature or vermine upon the earth, but hath its like in the seas, and that there is a Toad-fish, yet this Author avereth, that that fish never commeth neare the shore, but is constantly in the depth of the Ocean, as is the Sharke, the Flaile-fish , and others of that noxious kinde, and that he never saw or heard of any taken upon any Coast save one, which was in the yeare that *Nero* (that never-sufficiently detested Tyrant) was borne in, of which he hath this note, that *Monstrum praecessit monstro:* and plainly divines that its arrivall was ominous, as indeed all Histories doe with constant consent maintain and write, that all unusuall births either in men or bruit creatures, in sea or upon land, especially out of their seasons, have ever been the fore-runners and sad harbingers of great commotions and tumults in States and Kingdomes, if not mournfull Heraulds of utter desolation: Witnesse the Heifer calving of a Lambe upon the Altar in Jerusalem, (mentioned by *Josephus)* some halfe a yeare before the dismall sacking, firing, and final subversion of that beautifull and renonwned City by *Vespasian.* A mares foling of a colt with two heads at *Vitellius* (that beast) his entrance into the Roman Empire, who did much mischiefs in his wicked raigne. A Whales comming ashore at *Diepe,* a little before *Francis* the first was taken prisoner at the bataile of Pavia by *Charles* the first, Emperour of *Germanie,* and King of *Spaine.*

Fig 1 Toad-fish. Wikipedia Creative Commons

These unnaturall accidents though dumbe, do notwithstanding speake the supernaturall intentions and purposes of the Divine powers, chiefly when they meete just at that time when distractions, jars and distempers are a foote in a Common-weale or Kingdome: Messengers of Justice they were ever accounted;nay,they have without missing ever proved themselves to be the same.God in his mercie grant that this ugly monster may not for our sins prove the like to us, seeing the divers sins which are by divers Divines comprised in the nature of a toade,raigne, and have their swinge in our Nation. It is further observed by those that professe skill in Prognostication, that of how much the monster is of feature or fashion, hatefull and odious, so much it portends danger the more dreadfull and universall; God defend their observation may not hold in us, but surely a creature (if a creature we may call it, though truly it goeth something against the haire, considering its detestable ugly shape) I say a creature more displeasing, and at which humane blood may rise, I never saw with my eye, nor desire to see againe.

Lord we beseech thee turne thy back upon our sinnes,and thy favourable aspect upon our miseries, very likely with more haste then good speede to light upon us. Unite (good God) Head and Members, King and Parliament, encrease their loyall affections to him, his royall approbation to them and their proceedings, whence may proceede in our time His Majesties content and all our securities, and let all true-hearted,plaine-dealing, and plaine-meaning Subjects say *Amen*

A "Winged Toad" in Suffolk in 1662

The following letter, from Thomas Flatman, dated September 25th,1662,to his brother,is in the Bodleian Library,Oxford. I found it in the Early Modern Letters Online database, which I thoroughly recommend. The illustration on the following page is by myself. The transcript below is taken from the style the Bodleian Library sent to me,just as in the original:

Transcription of MS Rawl.letters 107, fol.204

Deare Brother

I have iust leysure enough to answere that part of yours wch concernes the newes of the Serpent- amongst us, I have not seene it myselfe but can name you 20 yt have all agreeing punctually in the[ir] relac[i]on & descripc[I] on of ye same; tis above a yard and an halfe long an head like a toade but very large a yellowish ring about ye neck 2 wings as broad as a mans hand like a Batts 4 yellowe short leggs like a ducke as bigg as a lusty mans Thigh the

9

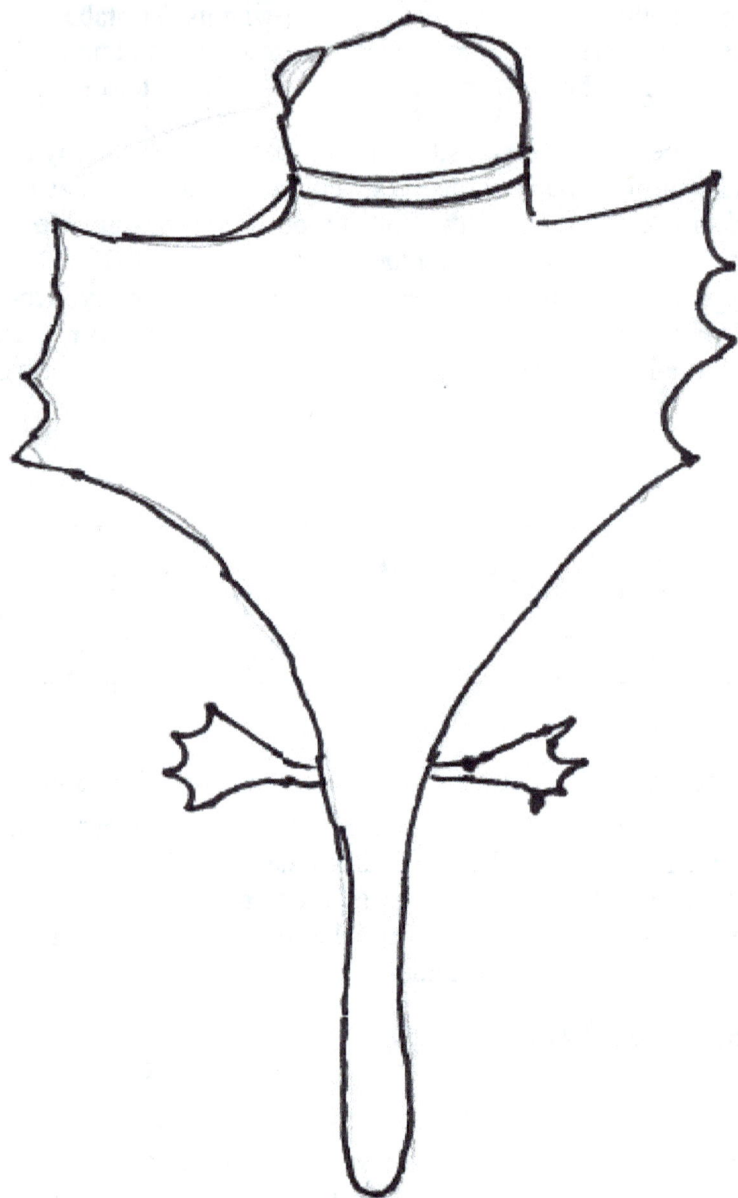

Fig 2 Possible appearance of Toad with wings

of 1662.

Front "legs" hidden beneath front
"wings".

Belly yellowe speckled with blacke spotts, head and back all covered with thick scales wch shine in the sunne reflect all manner
of coullers hee was seen eating a water henn is most often seene
before sunn rise in the morning and about noone when the Sunne
shines bright and hott. Heere is one affirmes that hee surprised
the Serpent one morning and being in a place where hee could
not retreate hee ris: & sprung att ye man but mis't him. The messenger expects my Letter. This with my Love is all the trouble
you rec[eave] frome Yo [S?][1]

                              Yo[ur] hartily affectionate Bro

                                   T Flatman

Mendham

Sept 25 1662 (1)

_____

[1] This is obscured in the text. It is possibly an abbreviation for
`Your Servant`.

Dr Karl Shuker in his 'From Flying Toads to Snakes With Wings' mentions a Welsh animal known as llamhigyn y dwr or water leaper, as follows:

" One of the most formidable water monsters documented in John Rhy's exhaustive two-volume opus *Celtic Folk Lore, Welsh and Manx* (1901) was the terrifying llamhigyn y dwr or water leaper, which inhabited lonely stretches of river in Wales and devoured any hapless sheep or other livestock venturing into its freshwater domain. Its body's shape recalled that of a huge toad, but there the resembled ended. Despite its name, the water-leaper lacked the toad's muscular, hopping hind legs. Instead, it had a tail- and a large pair of wings!

Nevertheless, this weird wonder did share one notable characteristic with the toad - a powerful voice, literally a hideous shriek, with which it deliberately frightened so thoroughly any unwary travellers seeking to cross the river that they would lose their footing and fall headlong into it. Once in the river, they would swiftly make a brief (and invariably fatal) acquaintance with its monstrous occupant. Happily, the loathsome llamhigyn y dwr does not appear to have been encountered lately" (2)

REFERENCES

1. Letter held in Bodleian Library also found on Early Modern Let - ters Online Database.

2. K.P.N. Shuker From Flying Snakes to Toads With Wings

    (1997) pp 105-106.

Frogs and Toads in Early Eighteenth Century Nottinghamshire

Whilst a patient in a Northampton hospital in 1997 I visited the town's public library and have had a copy of the following document ever since. It is an extract of pages in The Natural History of Northamptonshire (1712), by John Martin, that deal with frogs and toads:

Some Ingenious Gentlemen, that I have talk'd with here, are of Opinion that of another Animal of this Class, *viz.* Toads, there are really *Two* different Sorts, one of the Colour of the common Toad; but with a smoother Skin, for the most part, if not constantly, inhabiting the Waters, and watery Places; for which Reason they call it the *Water-Toad;* this Sort the very Worthy Mr. *Kirkham* of *Finshed* tells me has seen in Coitu with a fair green Frog: The other, of a dryer, rougher, and more husky Skin, which is the common Sort. Some others again incline to think that a Toad after all does not differ specially from a Frog, and that what is vulgarly call'd a Toad, is only an old overgrown Frog. I am of a different Opinion from these last; there being very small Toads, so I call them, they having the Colour of the larger ones, and the like progressive Motion: and a Toad having Faculty when provok'd of pissing or of spirting out a black and liquid Matter behind, that a Frog has not. But whether there are distinct Species of Toads with us, I am not so well assured, having never yet had the Hardiness of medling with them so far. That there are in the World different Species of Toads is certain: And the hairy one call'd the *Spanish Toad,* which was lately in the Physick Garden at Oxford, was probably a different Species from the common. But it must be granted that differences in Colour, and a greater or lesser Roughness of the Skin, such as in the Toads above described, may proceed from accidental Causes, and that Animals thus diversify'd, may nevertheless be of the same Species. And as to that call'd the *Water-Toad,* I am of Opinion, that 'tis only a blacker or darker colour'd Frog; the rather, for that I have in some places observ'd, particularly in a Pit or Well not very deep, by the Road-side below *Thrup-mandeville,* of Frogs a great Variety, as to Colour: some of a lively Green, others a Livid, others of them Yellow, and others in Colour exactly like a Toad; but in Magnitude, in Shape, and in Manner of their Motion, they agreed, and were all of them Frogs. The Toad-colour'd ones were as nimble Leapers as any of the rest; which Variety of Colour I suspect is ow-ing to the Differences in Sex, Age, Vigour, and the Places where they generally feed. 'Tis likely the Green and Yellow Frogs in the Well descended down thither from the Surface, and that those of a more sable Hue were bred below. It may by anyone be observed, that those found in green Herbage, that have a clearer sort of Feed-ing, are of a brighter, those in dirty Holes are of a darker Hue: And that the Male Frogs, about their engendering time, are as black al-most as Toads. (1)

Carl Marshall suggested, in an e-mail to me on September 5th 2012:

" Hi Ric,

All I can think of with this hairy toad is that it may have been the monotypic " Hairy" frog Trichobatrachus robustus from central Afri-ca, family: Arthroleptidae, that (I thought) hypothetically could have been unwittingly imported along with some unknown flora by Ox-ford Botanic Garden, survived the trip, was discovered and for some reason oddly identified as Spanish.

T.robustus also has a disturbing defence mechanism. When threat-ened they will actually break the nodule connection in their toes forcing the bones out, giving them makeshift claws. This singular activity was witnessed by the late naturalist and zookeeper Gerald Durrell. How macabre!!" (2)

REFERENCES

1. John Martin. The Natural History of Northamptonshire (1712)

2. E-mail from Carl Marshall to Richard Muirhead September 5th 2012

Chinese Coins
in Unexpected Places
Richard Muirhead and Bob Skinner

From time to time Chinese artefacts, especially coins have turned up thousands of miles from China`s shores, even as far aware as Ireland. [I`m not looking into the question of the Chinese porcelain seals that have turned up in Ireland as this has been covered elsewhere-R.] In March 2012 Bob Skinner, a Fortean from Surrey, did some research into Corliss` investigation into this matter and turned up the following:

In Corliss` " *Ancient Man- A Handbook of puzzling artefacts*" (1978) the following articles are reproduced:

pp. 429-30: " Chinese coin in Alaskan Burial" - Anon from Nature 46:574-5 (1892)

pp. 430 " Chinese coins in British Columbia"- James Deans;American Naturalist, 18:98-99 (1884)

A few articles re China and the Americas in early times:

p. 709 " China and the Maya" Anon; Nature, 133:68 (1934)

pp. 711-18 " Was Middle America peopled from Asia?" Edward S Mors; Popular Science Monthly, 54: 1-15, (Nov 1898)

p. 718 " The similarity of Chinese and Indian languages" Anon,Science, 62: sp x11, (Oct 9 1925).

Two or three similar articles are also found in Corliss` earlier "*Strange Artefacts- a Handbook of ancient Man*" vol M2 (1976)…

15

In Corliss' "*Ancient Man - A Handbook of puzzling artefacts*" (1978):

pp. 621-2 " Odd Old Stones" - Pete Pindar; Producers Review, 66:47, October 1976, which includes a paragraph on the find of an Egyptian Scarab in the early 1900s on a North Queensland cane farm. (1)

The October 7th 1944 edition of the Irish paper the Anglo-Celt reported that Master Jim Cleary of Teehill, Co. Monaghan, Ireland,unearthed a Chinese coin dated 1243. (2)

The Epoch Times web site posted the following story dated October 30th 2011 by Joan Delaney:

300-Year-Old Chinese Coin Found in North of Canada

A Chinese coin more than 300 years old has been found near a proposed mine site in Yukon in north of Canada. James Mooney, a cultural resource specialist with Ecofor Consulting Limited, spotted the coin while doing heritage impact assessment work for Western Copper and Gold Corporation.

"I was less than a metre from our archaeologist Kirby Booker when she turned over the first shovel of topsoil and I caught sight of something dangling from the turf. It was the coin - the neatest discovery I've ever been part of," says Mooney.

Minted between 1667 and 1671, the coin was found within the Selkirk First Nation traditional territory on the historic Dyea to Fort Selkirk trade route.

The coin adds to the body of evidence that the Chinese connected with Yukon First Nations through Russian and coastal Tlingit traders during the late 17th and 18th centuries and possibly as early as the 15th century, according to a release from Western Copper and Gold.

Fig 1 A map of Yukon territory

Canada

Wikipedia Creative Commons.

Although common along the northwest coast of present-day North
America, only three coins have been found in Yukon to date. The
coins are found with a square hole in the centre, but the one found
four by Ecofor has four additional small holes above each corner of
the central square.

"The extra holes could have been made in China; coins were some-times nailed to a gate, door, or ridgepole for good luck," says ,Moon-ey.

"Alternatively, First Nations might have made the extra holes to at-tach them to clothing. They used the coins as decoration or sewed them in layers like roofing shingles onto hide shirts to protect warri-ors from arrow impacts." The Russians traded items such as tobacco, tea, beads, firearms, iron implements, kettles, needles, clothing, and flour directly with the Tlingit in exchange for a variety of furs, which they traded to the Chinese in exchange for goods.

Mooney says the location of the find, on a promontory overlooking a river and creek tributary, is a likely place for a traveller to have rested or camped between Dyea, Alaska and Fort Selkirk in Yukon.

Although the coin was discovered in July, he says a fact-checking had to be done and information gathered before the find was announced publicly. The history of the coin is special in that it was a number six in a series of "poem coins" that were used as good luck charms dur-ing the reign of Emperor Kangxi of the Qing Dynasty.

Kangxi was renowned for his poetry. He was also associated with peace, prosperity, and longevity, so people gradually developed the custom of collecting a coin cast from each of 20 mints, putting them on a string and carrying them for good luck. The coins were placed in a certain order to create the poem.

Of the other two Chinese coins found in Yukon, one was minted 1724-1735, and the other, discovered back in 1993, is from between 1403 and 1424. The coin found in 1993 was discovered in a travel corridor near an overland gold rush trail by Beaver Creek. However , because it was found in an archaeological setting, it was likely brought into the interior before the Klondike Gold Rush.

So far I believe each of these three coins was found only with prehis-toric materials and no other historic materials, making them likely traded into the interior", says Mooney (3)

REFERENCES

1. E-mails from Bob Skinner and various books by W.Corliss

2. Anglo-Celt October 7th 1944

3. 300-Year-Old Chinese Coin Found In North of Canada. The Epoch Times http://www.theepochtimes.com/n2/world/300-year-old-chinese-coin-found-in-yukon-...

Fig 1   Wikipedia Creative Commons

Ancient Chinese Coins

Telling The Time from Cats Eyes

Richard Muirhead

The following story appeared in the **Grey River Argus** of June 11th 1904. This was a New Zealand newspaper.

WHAT`S THE TIME, PUSSY?

According to the French missionary Huc, no man needs a watch or a clock if he has the right kind of a cat. In certain parts of China they can tell the exact time of day or night by looking into a cat`s eyes. The pupil of the eye, assuming that the creature in question is just what it ought to be, gradually diminishes as noon approaches, until it loses completely its oval form, and becomes a thin perpendicular line. When that line is plumb it is 12 o`clock.

Then the pupil begins to grow very gradually, and finally becomes as big and as round as a marble. Then it is midnight. With patience,practice, and good mathematical perception, the happy possessor of a time-keeping cat can tell the hour of the day and of the night, because the thin perpendicular line which the pupil of the cat`s eye assumes at noon gives him a clear starting point.

The missionary discovered this valuable piece of feline peculiarity by pure accident. He noticed a little boy minding a calf, and asked him if he knew the time. The boy ran into the nearest hut, and came out with a big cat in his arms. "It`s just half-past eleven ," he shouted . And, running up to the missionary, he placed the cat`s face under Hue`s nose. Later on, when he got among his converts, he asked them to explain the mystery. They did so, and showed him some living specimens of the precious   timekeepers. (1)

This story actually dates back to Huc's *Chinese Empire* (1854) which says:(according to a posting on Dr Beachcombing's site www.strangehistory.net - Cat Clocks - No Really! February 28, 2012:)

"One day when we went to pay a visit to some families of Chinese Christian peasants, we met, near a farm, a young lad who was taking a buffalo to graze along our path. We asked him carelessly, as we passed, whether it was yet noon. The child raised his head to look at the sun, but it was hidden behind thick clouds, and he could read no answer there. `The sky is so cloudy`, said he, `but wait a moment`; and with these words he ran towards the farm, and came back a few minutes afterwards with a cat in his arms. `Look here`, said he, `it is not noon yet`; and he showed us the cat's eyes, but pushing up the lids with his hands. We looked at the child with surprise, but he was evidently in earnest. `Very well`, said we, `thank you` and we continued on our way...

As soon as we reached the farm...we made haste to ask our Christian friends whether they could tell the clock by looking into a cat's eyes. They seemed surprised at the question; but as there was no danger in confessing to them our ignorance of the properties of a cat's eyes, we related what had just taken place. That was all that was necessary; our complaisant neophytes immediately gave chase to all the cats in the neighbourhood...

They brought us three or four, and explained in what manner they might be made use of for watches. They pointed out that the pupils of their eyes went on constantly growing narrower until twelve o'clock, when they became like a fine line, as thin as a hair, drawn perpendicularly across the eye, and that after twelve the dilation recommenced."(2)

REFERENCES

1. Grey River Argus June 11th 1904

2. Cat Clocks - No Really! February 11th 1904
www.strangehistory.net

A Giant Crocodile Ballad

I found the following interesting ballad on the National Library of Scotland's web site (1).It's just a piece of doggerel really.

This most wonderful song came out of the Poet's Box, and can only be had there for the price of One Penny

AIR - End for End Jack

Come list, ye landsmen, unto me,

To tell you the truth I'm bound,

Of what happened me whilst I was at sea,

And the wonders there I found.

Shipwreck'd was I, just off Peru,

Scarce half a league from shore,

So, resolved was I, to have a cruise,

The country to explore.

Oh! ri tol de rol, &c

I scarcely there had scudded out,

When close long side the ocean,

I saw something move, which, at the first, I thought,

Was all the world in motion.

I quickly bore longside of it

And found 'twas a crocodile,

And from his nose to the tip of his tail,

It measured five hundred miles.

Oh! ri tol de rol, &c

This crocodile, I could plainly see,

Was not of the common race;

I was obliged to climb a very high tree

To get a sight of his face!

My eyes! When he did ope' his jaws -

Now, perhaps you'll think it a lie,

'Twas above the clouds for miles three score,

And his nose quite touched the sky.

Oh! ri tol de rol &c

I being aloft, the sea ran high,

It blew a gale from the south -

Lost my hold, and away did fly

Into this crocodile's mouth;

He quickly closed his jaws on me -

He quickly closed his jaws on me -

Thinking to grab a victim -

But I ran down his throat, dy`e see,

And, (damme) that`s the way I trick`d him

Oh! ri tol de rol, &c

I travelled on for a month or two,

`Til I got in his maw,

Where I found of rum kegs not a few,

With plenty of bullocks in straw!

Of life I banished all its cares -

For, of grub I was not stinted;

In this crocodile ten years I lived,

And was jolly well contented.

Oh! ri tol de rol &c

This crocodile being very old,

Alas! One day he died;

He was full five years a-getting cold,

He was so long and wide!

His skin was five miles   thick, I`m sure,

25

Or somewhere there about,

For I was full six months, or more,

Cutting a tunnel to get out.

Oh! ri tol de rol, &c

So, now, you see me safe on land,

Determined no more to roam;

In a ship that I passed I got a berth,

So you see me safe at home.

But if my story you should doubt -

If ever you travel the Nile,

Just where he fell you'll find the shell

Of this rummy crocodile.

Oh! ri tol de rol &c

R E F E R E N C E

1.http://digital.nls.uk/broadsides/broadside.cfm/id/16426/criteria/Cr
ocodile

A Wonder Boy

This strange story was found at the same time as the Crocodile ballad. I don`t know whether it`s well known or not, but I thought I`d
include it in *Flying Snake* anyway.

"Wonder of Wonders, or the Speech of a child born near Edinburgh
on Thursday the 15th of March 1770 as delivered ten minutes after
it came into the world.

In all ages scarce such an instance is to be parrelle`d as the present
incident, which whatever the public may think is assertain`d for
truth.

On Wednesday the 14th of March a farmer`s wife near Edinburgh
being within a short time of her delivery, being weary`d and went
to bed, and after an hour`s sleep arose and said to her husband,

My dear, I have had a sweet sleep and has but few hours till I bring
forth a Son, who will tell what shall shortly happen.

Accordingly, on Thursday following, the woman was happily delivered of a son who also as was born spoke the following words.

This year will be a bloody year, there will be war proclaim`d before
the end of it, but the christian powers shall obtain the victory;
Highest powers shall be insulted; Wilkes`s party shall be sadly
dismay`d on account of their irregular conduct; while the SCOTS
shall be prais`d for loyalty to their King; London, London, the seat
of our rightful Sovereign, the Thistle is sharp, but never sting`d
you;But the charming rose, in its full bloom, so cheers the hearts of
the 45th party, that the Honourable Ancient Thistle appear, they
mind nothing but revelling; but says the child, or the end of the

the year 1770 forty five blue bonnets will drive 300 Freeholders to flight, Wilkes, the unhappy instrument of our disturbance shall with his party be brought to disgrace and be oblig'd to bow with reverence, to them whom they ineffectually endeavour'd to afront.

So spoke the child, and immediately expired. (1)

REFERENCE

1. http://digital.nls.uk/broadsides/broadside.cfm/id/16824

CARL`S COGITATIONS: GREAT SNAKES OF THE WORLD PART TWO

" As to what is meant by weird - and of course weirdness is by no means confined to horror - I should say that the real criterion is a strong impression of the suspension of natural laws or the presence of unseen worlds or the forces close to hand." H.P. Lovecraft in a letter to Wilfred Blanch Talman August 24th 1926

Fig 1   Head of an European Grass Snake . Natrix natrix.

Collection of Carl Marshall

North Africa. Unknown crested snake:

Very large crested snakes are reported from eastern Morocco to Tunisia (also see Crowing Crested Cobra - Central/Western Africa), could they somehow be surviving pythons? or something completely new! The crests reported may just be where old or unhealthy individuals are struggling to slough (shed) their skins completely with some skin retained around the neck and head and building up over successive moults until they eventually resemble very old crested snakes. This author has researched these cryptids in Morocco and found they are still reported.

Central/Western Africa.

Rock python:

The rock python (*Python sebae*) is a large species from sub-Saharan Africa. There are two sub-species, one from Central/Western Africa and the other from South Africa.

P. sebae is Africa's largest species of snake with specimens reported (but not confirmed) of individuals reaching and exceeding 20 ft (6 metres+). Although these estimations haven't been confirmed they are considered, by most to be quite possible.

This species lives in a variety of habitats near water from forest to near desert and they have been known to eat antelope and even crocodiles which they kill via constriction.

This species has killed humans and actually eaten children;

A Uganda newspaper reported in 1951 that a 13 year old child had been killed and consumed by this species. Although the child was eaten he was later regurgitated.

In 2002 a 10 year old child was confirmed to have been eaten in South Africa.

Could this species or a similar one (albeit a larger type) be responsible for the giant snakes reported in much of central/western Africa?

Crowing Crested Cobra:

Stemming from folklore similar to the Basilisk or Cockatrice, the Crowing Crested Cobra is considered to be a large snake but with three very significant differences

Fig 2 Drawing of crested snake by Maureen
Ashfield (modified spectacled cobra Naja naja.)

This mystery serpent is said to not only crow like a cockerel but to also have a blood red crest on its head and wattles, just like a chicken, sometimes this mystery snake sometimes bears a traditional Cobra's hood however most often this is absent. It is sometimes said, simply to be a snake with a crest of feathers.

Like the fictional cockatrice this snake is said to come from an egg laid by a chicken and hatched by a toad. This creature has a huge range, or shall we say the folklore that surrounds it does, not only is it reported by natives over much of the Dark continent, and Asia but there is also a much smaller neotropical version reported from the Caribbean typically on Jamaica. The Crowing Crested Cobra has also been spotted by many respectable western explorers and travellers.

A medical doctor reported seeing one of these unusual snakes in 1829 on the island of Jamaica.

A snake with wattles was shot and killed in 1850, also on Jamaica.

Reported over much of East and central Africa and known by many names such as the Bubu (Shupanga), Hongo (Chi-ngindo), Songo (Chi-yao) and to the babwe natives of Zimbabwe; the mbobo. Richard Freeman of the CFZ recently came back from an expedition to the Garo hills in search of the Indian Yeti or the mandeburung. Richard came back with the tail of a giant crested snake called the Sankuni that he likens to the Naga - a large crested serpent from Thailand.

The African species is reported to measure up to 20ft (6 metres) just larger than the largest King cobra's (*Ophiophagus hannah*) and if discovered to be a real flesh and blood animal would become the largest venomous species of snake in the world. The Crowing Crested Cobra is reported to be a brown or greyish black with a scarlet head although Its new world counterpart is described as a dull yellowish brown with dark spots and only reaching a maximum of 4ft. For an extensive, detailed article on this creature I highly recommend reading Dr Karl Shuker's Extraordinary Animals Revisited (CFZ Press 2007).

Observation of a U F O in Sutton, Cheshire, Spring 2011

Trevor Bailey

My girlfriend and I were driving up to my parent`s house , in Sutton, (just outside Macclesfield), in our car, and were just nearing tall houses on the left-hand side of Walker Lane. (Which used to be an old ragged school in Victorian Times.) When I looked over to my left to open fields (this would be in the afternoon, on a clear, bright day), and I was amazed to see the most awesome sight, as there hovering off the ground was the strangest sight I have ever seen, a definite U.F.O sighting; as a black, disc shaped craft floated in one position, not too far from us.

The material it was made from looked like shiny, black plastic, and in the centre of the machine was a round , deep recess with a helmeted figure sitting in it. (Also dressed in a shiny, black material).

I wished that I`d have had my camera with me to take a photograph of it before it disappeared . I wonder what would have come out on the developed photograph? ( Sadly I will never know!) It was an experience of a life time, and an event I will never forget. I`m afraid I do now believe in U.F.O `s.

Fig 1   Trevor Bailey's own impression of the U.F.O he
saw in Sutton, Cheshire.

Purists will realise that as U is a vowel, I should have written "an U.F.O.", However, I thought "a U.F.O." sounded
better.

Words from The Koran on an Indian Ocean fish

MYSTERIOUS FISH WITH STRANGE ARABIC

INSCRIPTION

Wide World Magazine Late 1917/Early 1918

The first sentence says: "The accompanying photograph (see page 37-Rich) depicts a remarkable fish which was recently caught at Zanzibar with strange Arabic inscriptions upon its tail [it] appears that the fish was not of a large haul, and was taken by a single fisherman, who brought it to the fish-market. There it remained for some time, having no purchaser as it was one that had never been seen before in those waters. Finally an Indian of the sect called "Memon"bought it, and, on the strange markings being noticed, it was taken to a well known Arab scholar, who deciphered the inscription. It was afterwards sent to the Sultan, who also recognized the wording. That night, three thousand rupees were offered for the fish and refused, and on the following day five thousand rupees were refused. The original price paid for the fish was a penny, and it was eventually decided to have it preserved. For this purpose it was taken to the Government Laboratory, where it was treated with formalin. It has since been placed on public exhibition. There are two distinct inscriptions on the tail, one on each side. One reads " The work of God", and the other, "God alone". There is no suspicion of anything in the nature of a fake about the matter, and the mystery is so complete that no explanation of the strange phenomenon is forthcoming. The Arabic lettering is perfectly plain, and the discovery has caused wonderment throughout the Mohammedan community of Zanzibar. They declare it portends something about to happen, possibly the end of the war.

the end of the war. In the course of conversation with a prominent East African official we gather that the only previous case of a similar nature known occurred in Tibet, where certain Hindustani wording appeared on the leaves of a tree. The whole of the circumstances are certainly very strange. (1)

On May 19ᵗʰ 2006 the BBC News reported : Kenya's `Koranic Fish` Recovered - A fish with markings that resembled a Koranic text has been found by Kenyan officials after vanishing from the fisheries office where it was stored.

The tuna fish, which had provoked intense interest from Muslims, had apparently been stolen by people posing as National Museum officials.

The fish was found at the shop where it had first come to public attention.

The fish was being studied to find out if the Arabic inscription "You are the best provider" was natural or a hoax.

Sceptics say the writing was the work of someone who caught the fish and then threw it back into the sea.

But others say this would be impossible, and local imams are said to be talking in the mosques about the fish.

Heritage

The owner of the Takaungu Fish Shop in Mombasa's old town is being questioned by police, who are also seeking another man in connection with the alleged theft.

The shop-keeper said a man had brought the fish to him for preservation.

Over the weekend, people thronged to the shop after the owner noticed the tuna fish's remarkable markings.It had been caught by fisherman Said Ali at the end of last week at Vanga, a small fishing port on the Kenyan coast, 50km south of Mombasa. For safekeeping, the 2.5kg (five pound)fish was moved to the fisheries department. After being asked by Muslim leaders in Kenya, Kenya's National Museum had offered to take custody of the fish and preserve it for the country's heritage.The reported theft followed numerous attempts by locals and Muslim scholars to buy the mysterious fish. An official at the fisheries department in Mombassa said someone had even offered to pay as much as $150. Under normal circumstances the fish would fetch not more than $6. (2)

Fig 1 The 1917 fish with the Koranic inscription.

REFERENCES

1. Wide World Magazine Late 1917/Early 1918

2. BBC News web site May 19th 2006

This Interview with Anthony Quiney (A) ex game keeper of the Ragley estate concerning big cats roaming the local area was conducted by Carl and John Marshall . The transcript is accurate, which represents Anthony's local terminology.

Interviewees : Carl Marshall (C)

John Marshall (J)

Permission was asked and given for Anthony Quiney's name to be disclosed during this interview.

Interview started at 2:10 pm

The following interview starts abruptly because Anthony started his story before I (Carl) had got the Dictaphone ready and when recording he started where he left of.

A: Yea, so there was this chap Quinton Rose, and he came up from London Zoo with viles of leopard pee to lure one in, and baited areas, put up cameras stuff like that. He showed me a map of all over the country where unusual species had been seen, there was a Wolverine up in the north somewhere, up on the moors that was.

C : Really that's interesting.

A : Yea, well because of this dangerous animals act people were just chucking them out, you know.

C : I certainly do.

A : And that's where this leopard, well pair of leopards (comes into the story) and there was one over at Hanly Swan and this was a male, and the female was here and her area goes from say Ragley was the start, she goes up the Lenches, she goes round to Flyford Flavel right over towards Worcester, Feckenham all that area, she's been seen up Marlcliff. Now about twelve years ago a chap was combining at Salford lodge now he said he saw her coming out of the corn back towards the wood and she had two cubs with her and that Quinton Rose he went out and interviewed this chap, but I know a lot of people who've seen her, Iv'e never seen her though but I know a lot who have, including my wife. C: Oh really, OK. A: Could I maybe talk to her about it sometime?

A : Sure. Oh there was this young lad kicking a ball over at Cold Comfort farm, he was about twelve years old, anyway he stopped kicking the ball and was watching a dove which had landed on the track, you know a farm track. He was just stood watching this dove when this leopard jumped through the fence grabbed the dove and turned, and she was only about three to four yards away from him, starred at him for a second then went off with the dove. He said she had grey hairs around her ears which could imply she's getting on a bit, you know quite old and he was on about the length of the tail and everything, you know he described her really well, and again this Quinton Rose interviewed him like.

C : Right OK.

A : Yea.

C : And what year was this?

A : Well I'm going back about twelve years now I would think so were are we know?

J : That would be about 1999.

A : Yea.

C : Its a real shame you haven't seen her with your own eyes.

A : Yea I would love to have seen her. But like I say I know a lot who have, Lennie Quiney (no relation) down at the saw mill he's seen her and both of the herdsmen who used to be up Ragley they seen her as well.

C : I was originally told that these animals was shot and were buried at Ragley is this true?

A : No.

C : OK so we will forget that then.

A : You know, a friend of mine shot a civet.

C : Oh really that's interesting

A : Yea.

C : Back on the leopards for now, do you know how big the cubs were?

A : No, you will have to speak to Graham Mills, he was on the combine, as far as I know he said all he saw was an adult come out and there was two cubs behind her, you know they came out of the corn.

C : And did he live local?

A : Yea Abbott Salford.

C : OK did you find any animal kills out there or any tracks or scat?

A : We did have a dead pit which they could have in those days, this was up at Ragley, and there was several calves took out of there and taken into the woods. But I said to Quinton Rose " I'm surprised we haven't found any kills up trees " and he said They only do that on the African

39

Fig 1

Mr and Mrs Quiney

they only do that on the African savannah where the leopards are
not the top predator the lion is."C: Yea . A: The lion would just
take their food you know. C: Can you tell me of any other unusual
species you have seen or heard during your career as a Game keeper? A: Yea a friend of mine shot a civet cat, that was over at
Wellesbourne that was. C: Right OK. A : Yea so he shot a civet,
and there was a skunk found and taken over to the Vale Wildlife
Park that was found somewhere in the Vale.
C : Yes I've heard about this.
A : But that's not so long ago.
J : Skunks are also commonly for sale so they breed them in the
country I bet there is a lot of escapes. You see them in the Cage and
Aviary magazine all the time.

A : I remember years ago going to a Civil Engineering show, it was in Hertfordshire I think, no it was Bedfordshire and there was a load of wallaby's in the next field over which had escaped from some wildlife park and were breeding quite well.

C : Yes there are lots of reports of wallaby's over the country.

A : Yea, they have been seen on Cannock Chase

C : Cannock chase does seems to be a bit of a hotspot.

A : We used to get a lot of different birds on Ragley, we've had hoopoe's there. Bee eaters, yea bee eaters have been seen many times at Hillers car park and one time a very large unknown bird of prey was seen soaring there that could not be readily identified by the ornithologists there.

C : I am not familiar with the hoopoe.

J : Also called the muck bird, it comes from Africa has a big crest on its head and it turns over the soil looking for grubs that's why they call it the muck bird.

C : Thank you. Going back to the big cats briefly what colour was the adult female.

A : It was all black.

C : Melanistic, thankyou.

A : Evidently people who have seen her up close said they could still see brown spots through the black, you know so if you see her a few meters away she would look black but if you got up close you could still see spots. That's what people have told me anyway. remember I have never seen it.

C : Any idea what colour the cubs were?

A : No like I said you will have to talk to Graham Mills. I tell you what I will go and have a word with him and if its ok I will take you over to meet him.

C : That would be great.

A : Yea I think it was Graham in the combine but if it wasn't he was in the tractor moving corn, but who ever was in the combine was the one who saw it so I will talk to Graham and find out who was driving the combine and they will probably be able to tell you more.

The most regular place it was spotted was Rous Lench Court, it is an old Tudor court and they have got a very big yew hedge there and at the bottom of this yew hedge there was an area with a build up of yew leaves and she used to sleep there quite a bit.

C : OK that's interesting. Do you have any ideas what it was mainly feeding on?

A : Muntjac, definitely muntjac.

C : OK.

J : There was a lot of muntjac in Ragley wasn't there?

A : Oh yea, the area she covered was covered with Muntjac. But she would go for anything though you know.

C : Yea.

TEA BREAK.

(Anthony carried on the interview where we left off)

A : And there was those calves out the dead pit and she dragged them out and over a fence into the wood and they were found skinned out which is usually a cat isn't it?

C : Yes quite possibly.

A : That chap Quinton Rose, he's is dead now, he was only thirty seven. Yea he went to the dentist to have a tooth out and he got septicae-mia and was taken into hospital seriously ill and when his father came to visit he didn't recognise him his face had swollen up that much you know, his head, his shoulders, everything. Yea and he died from that.

C : Yes I have heard of him before.

A : He made a trap, a live catch trap, it was such a good design it is used a lot now in Canada for catching all manner of animals because it comes in different sizes, so he actualy designed it. You know it was brilliant, he showed me, he put his hand in and it could hold him and didn't mark him.

C : Anthony, this story may get published so do you mind your name being used?

A : No, not at all.

C : Fantastic

J : Did you experience anything unusual while you were a keeper at Charlbury?

A : No nothing, we used to get deer there like sika, nothing that you might call exotic.

A : Quentin showed me a map of the British isles with unusual species on it, there was a bear right up north, there was all sorts of things. The police got involved when my friend shot the civet cat and he had to dig it up so it could be verified, he actually buried it.

C : Do you know of anyone seeing big cats in the area recently?

A : I will find out a bit more about it, I haven't heard of anyone seeing for years but I will ask around. If it was the same individual it would be a hell of an age now but maybe the cubs are still about.

C : And did you see the dead calves yourself?

A : Yea I saw the calves and they were big they weren't small calves they were quite big.

C : Also with cat kills the cat usually eats the prey from the back-end upwards is that how they were?

A : That's how it was.

C : It would be interesting to find out what colour the cubs were.

A : Well I will go speak to Graham mills see what he says. I mean I cant remember if he was driving the combine or moving the corn on the tractor. It was the combine driver who saw spotted them.

C : So where was this sighting again?

A : It was over at Salford lodge in Bevington Waste so its not to far away.

C : Have you heard of any surviving pine martens in this area?

A : No, polecats are fairly common but these are really just feral ferrets with traditional polecat markings. Pine martens are not impossible, they are in Ryadar in Wales and that's not to far from Worcester is it.

C : No.

A : I have seen many dead polecats on the roads.

C : So when this big cat was around there were many other corroborating sightings to confirm it?

A : Yea they were seen all the time by dog walkers etc

C : OK.

A : I know lots from the Ragley estate who have seen it, weather they are still working there I don't know, but Lenny is, Lenny's still there so if you want to have a chat with him.

C : What's the current head keeper like?

A : Paul.

C: Yea. A: I don`t really know. I know of him. C: Do you think he would let me have a look around?

A : That I don't know.

A : You see when Quinton came it was Lord Heartford who rang me up and said this chap was coming because what he wanted to do was take a DNA sample of it to see where it was released from. If you kept them in captivity you had to have a blood sample taken of the cat so it could be checked out in case it attacked anybody, so he was trying to get a sample so whoever released it could then be prosecuted.

A : The hunt were in Grafton wood over near Flyford Flavel and they put it up there, that was the West Warwick's , and a lot of the huntsmen saw it and there dogs wouldn't go near it. So if you talk to members of the local hunting fraternity they might be able to help you. Iv'e also heard of farmers out that way who have sheep go missing but what it was, I don't know.

C : Did you see any tracks on the Ragley estate?

A : Well if you look at mud, mud seems to enlarge most tracks any-way so if someone goes down with a big dog you know, but cat prints don't have the claws.

Iv'e rang up Quinton at times and told him of these tracks in so and so gateway and he just said no that's a dog print or whatever. But he actually sat up at Rous Lench Court and filmed it, yea he set up cameras there. They were good enough to let him stay in the house. There was this old chap there who used to see her regular on his lawn as he was having his breakfast. So Quinton stayed a few nights and actually filmed it there, so it is on film.

I think the powers that be know they are here but don't want to alarm the public. There was one chap who was ex army who took a shot at it in Inkborough church yard, he was roost shooting - wait-ing for pigeons coming in to roost , and he walked back through the church yard the leopard saw him, well they came upon each other. The leopard went to get away from him and in such a rush jumped into a tomb stone, came back at him, took a swipe at him, and he took a shot at it. He had on a waxy coat and the claw marks went through the waxy and into his shirt. Maybe If you go over Inkbor-ough you will find out more about this.

C : I take it he missed? A: Yea

C : So Quentin also believed that a female was in this area?

44

A: Yea that's right.It was definetely the female who was here the male was over at Hanley Swan which is Worcester way. C: OK. A: I am only going by what that Quentin Rose told me. Whatever dragged those calves out of dead pit was obviously very strong.C: Well thanks very much for your time Anthony this has been great. A: No problem, I will speak to Graham Mills and get more information on the cubs.

C: That would be fantastic

A: OK

Interview terminated at 3:00 p.m

NOTES: When I first heard of these encounters with the 'Beast of Burford'...I thought to myself I can't believe anyone hasn't looked into this story before. After all, this tale in its preliminary condition had some very interesting twists to it. First of all the adult female that was supposedly haunting the Ragley estate was originally described to me as " A normal phase leopard " and three cubs of unknown colour were also described. The three cubs soon became two and appeared to be the main subject of this story because all "three" were apparently shot in a corn field on the outskirts of the estate and buried in shallow graves within the grounds of the estate. I thought I had stumbled on a big cat story well worthy of investigation, imagine if I could locate what was left of these corpses and get a DNA sample, what a story this would be. Of course this was not the case as you will read later. I previously believed it had happened around 1981 as it turned out I was completely wrong.
Anthony Quiney is a very interesting man, he has very good knowledge of the local wildlife in the area having spent many years as a game keeper being out in the thick of it so to speak looking after game animals by preventing predators attacking them. As far fetched as it sounds Anthony can tell by the calls of blackbirds what predators are in the immediate area, a talent not to be underestimated with his occupation.
However Anthony (as did I) made a few mistakes during the interview which I am going to address below.

Comments: So much for the normal phase Leopard.(spotted)
After interviewing Anthony the normal spotted leopard soon
disappeared into the shadows only to be replaced by the more
traditional black panther. I was disappointed because as far as I
am aware there have been no, or very few, reports of normal
leopards in the United Kingdom and there are literally thousands of melanistic (black panther) reports, which can be
blurred into a modern feline comparison of the ancient black
dog legends of England such as the infamous Black Shuck.

The Dead Cubs.
My reference to the dead cubs was based on information that
was inaccurate and from a different source. The only part of
Anthony's story that mentions shooting leopards is when the
adult female was shot at in Inkborough church yard by the ex
army soldier roost shooting. I am going to look into this attempted shooting in more detail soon but for know I am inclined
to have doubts about the legitimacy of this part of the story.
It all happened in 1981.
According to Anthony's testimony and my own post interview
research this all happened between 1997 and 1999.

Quentin Rose.
Anthony stated in the interview that Quentin Rose died at the
age of thirty seven where as my research imply's he was actually forty six. I haven't been able to find out the exact cause of his
death but according to Neil Arnold's site from the CFZ (Centre
for Fortean Zoology), Quentin Rose was a chronic diabetic and
this probably contributed to his untimely demise. Neil has currently sent me an email informing me that he has more information on this subject ASAP. I await this email in anticipation.
It seems that Quentin investigated the Ragley beast with his
good friend Chris Bosley and together they apparently set up a
water proof tape recorder and laid down lion urine and dung to
try and draw the leopard in. I don't know quite how useful this
method would have been as I believe this would not have attracted a leopard but rather discouraged it even if the leopard
had no memory of lions   46

due to a life in complete captivity. The strong smell charged with pheromones would have told the leopard that there was a larger predatory Felid in the area and to avoid it. These messages would even tell the leopard the condition of the lion.

Towards the latter part of his life Quentin dedicated his life to his own new design of animal trap (the rose cuff), that caught the animal but did not harm it and was useful as it could be used on a variety of mammals. Sadly Quentin died in 2002 before he saw his dream made reality but thankfully his father George Rose picked up the torch and continued with his son's noble work. As Anthony said these traps are now very successful in Canada.

The Woman who gave birth to rabbits - the Mary Toft case

Oll Lewis

OLL LEWIS: Crypto Cons - More Rabbit Than Sainsbury's.
Part One April 8th 2011

In the 18th century very little was known about genetics, indeed it was not until the 19th century that Gregor Mendel came up with the concept of genes and even then his work was largely unknown for decades, and doctors, scientists and lay-people had no real idea how it was that humans came out of other humans in vaguely human shapes. To most people this didn't really matter to be honest, they just plopped out looking all "humany" and that was an end to it, after all logic would dictate that because you don't get animals giving birth to humans why would a human give birth to an animal? Following on from that logic this was proof to some of the unchanging and well ordered nature of the natural world, imagine the trouble you would have if fish gave birth to humans! Why, the person would drown! The general lack of baby corpses littering rivers was proof enough of the well designed order of nature.

Except nature, and humanity wasn't always well ordered, sometimes there are birth defects or complications during pregnancies. These were, in the 18th century often blamed on some animal having interfered with the pregnancy in some manner. The more uncharitable doctors, midwives and gossips might insinuate that the mother had conceived the child from having had congress with an animal and the more tactful might have suggested that this was as a result of the mother having been startled or scared

by a similar animal during the pregnancy causing the baby to take
on the look or traits of that animal .This bizarre superstition persisted into Victorian times among the great uneducated masses of
the city slums of London as it was used by sideshow exhibitors as
the explanation for why Joseph "The Elephant Man" Merrick
looked the way he did.

These hoaky home-spun theories would be pushed to the limit by a
strange case in Godalming, Surrey in 1726. Mary Toft the wife of
a clothing salesman, miscarried something that had the appearance
of a rabbit but with exterior lungs and heart. About 14 days later
Mary was said to have given birth to a live rabbit, followed by several more over the subsequent hours and days. None of the bunnies
survived for more than a few minutes, but several people are said
to have seen the births and the rabbits afterwards.

Reports of the event were published in Mist's Weekly Journal and
eventually reached the ear of King George I as a result. The king,
who was very intrigued by the reports, was to send his own investigators, his surgeon Nathanial St Andre and the secretary to the
Prince of Wales Samuel Molyneux, to Godalming to find out more
about the case. It turned out the event had not been a one off and
that Mary was still giving birth to rabbits and bits of other animals.
One midwife, John Howard, who had initially been sceptical of the
claims, had supervised at the birth of 3 cats legs, one rabbits leg,
the guts of an animal Howard presumed was a cat and the back-
bone of an eel. The explanation for the rabbit and cat births was
thought to be that Mary had dreamed about or strongly desired
each creature during her pregnancy so her baby had turned into
them. Shortly after the arrival of the kings investigators Mary gave
birth to the torso of a rabbit and St Andre examined her determining that the rabbits had indeed come from her womb. Later that
evening Mary gave birth to another rabbit torso in their absence
and a head in the presence of the two men. The investigators left
either wholly convinced, or perhaps in on the scam although neither actually confessed to it, and submitted a report to the king.

Deciding that all this was so compelling that it needed further study the King then sent another surgeon, Cyriacus Ahlers to investigate Mary Toft and the ever present midwife John Howard. Ahlers was a lot more sceptical of the claims than St Andre and Molyneux had been especially because, when he arrived Ahlers was showing no signs of pregnancy but proceeded to plop out a few bunnies for him on queue. Ahlers noticed that prior to these births Mary had been holding her legs together as if to prevent something falling out and that John Howard insisted in delivering the bunnies with no interference from Ahlers and Mary would cry out in pain whenever Ahlers came too near. When he left he pretended to be convinced in order to get his hands on some of the birthed bunny bits to study further. Ahlers determined that they had been cut with a knife and found straw and grain in their digestive tract and faeces. All was not looking good for Toft and Howard's story and things were also looking bad for St Andre as well who had endorsed the claims. Howard somehow caught wind of what was happening, probably via St Andre, and wrote a letter requesting the immediate return of specimens. St Andre returned to Surrey to see Howard and was handed two more rabbits that Mary had supposedly given birth to. St Andre also used the trip to collect affidavits from all concerned just so he could use them to cast doubt on Ahlders' version of events should he ever need to. St Andre then gave the king an anatomical demonstration of the births which led to the king asking for Mary Toft to be brought to London for more examination.

There were two comments: The first from Dale Drinnon:

My mother absolutely believed in the story of the woman that gave birth to rabbits: I had to break the news to her later that it was a hoax when I learned of the details. My mother also absolutely believed human males had a penis bone and they could be "ruined" by somebody breaking the bone-a horror story she used to admonish us with. I later found out that was not true either-evidently her uncle had told her these stories and she grew up believing him.

Best Wishes, Dale D.

The second from Richard Freeman:

The idea of imprinting the features of an animal that scared the mother during pregnancy. Lingered much longer than this. My grandad was convinced that a man he saw riding a bike on a regular basis in his youth had been imprinted with pig features when a swine had scared his mother. The unfortunate fellow had one large, pig like ear, a tusk and a snout like nose as well as a ruddy, porcine completion.

Part Two.   April 10ᵗʰ 2011

You would think that being moved to London at the request and ex-pense of the King of England would have made Mary Toft and her col-laborators think twice about organizing more human-rabbit births. It was probably organized and agreed to in the first place as an exercise to call her bluff; surely, if this was all a hoax and they had any common sense, she, her husband Joshua and midwife John Howard would see sense and abandon the charade before they got in any deeper. Unfortu-nately though they were not really that intelligent and like most stupid people who get into similar circumstances they were probably busy congratulating themselves on how they had been clever enough to fool all these big city types with their fancy qualifications and thought them-selves to be invincible. The trouble was that the Toft and Howard were not fooling everyone, in fact it was probably only the first surgeon who had seen them, Nathanial St Andre, and the secretary to the Prince of Wales, Samuel Molyneux, who had wholly convinced by the validity of Mary Toft `s rabbit births. St Andre, the King's surgeon, had foolishly staked his reputation on Mary's unusual births and issued challenges that anyone who did not believe him could visit Mary and witness a birth in person.

St Andre's challenge was one thing when Mary was living out in the sticks in Surrey, only one surgeon, Cyriacus Ahlers, who had like St Andre also been sent by the King, bothered making the trip. Ahlers found evidence of the rabbits and other animal parts Mary had been giv-ing birth to having been cut by knives and found straw in their drop-pings so was less than impressed by the evidence. However, when Mary was moved to the bustling metropolis of London, right slap bang in the middle of the enlightenment it put her within easy reach of most of England's reputable surgeons and gentlemen scientists so visits and investigations became more frequent. Naturally Mary obliged her visitors by conveniently plopping out bits of meat in their presence every time, despite usually not appearing to be pregnant.

One such visitor was Richard Manningham, who had gone to collect Mary Toft from Surrey with St Andre. Manningham was completely unconvinced by Mary and identified one of the bits she had plopped out for him as a fully grown pigs bladder that still contained urine. St Andre was able to convince some of the London set with Mary's displays though, John Maubray, one of the leaders of the male midwife movement, gladly jumped on the bunny birth bandwagon as this vindicated his theory of "sooterkin". According to Maubray, and several others at the time, Mary's strange rabbit births were small creatures named sooterkin, formed in the womb as a result of over familiarity with household pets. A more respected midwife than Maubray and expert in female anatomy, James Douglas, was often invited to view a birth by St Andre but felt sure that the births were a hoax.

While in London it was decided to put Mary Toft under constant supervision and during this time, whenever she had a visitor that wanted to see her giving birth to a rabbit she would go into labour but produce nothing.

The endgame for Mary and her collaborators began when Thomas Onslow started to investigate the affair. Onslow had Mary's husband Joshua, who had been more or less ignored by most people investigating the case, followed. Joshua was caught red handed buying rabbits for use in the hoax. Upon hearing the news that Joshua had been caught buying rabbits one of the porters that was charged with looking after Mary also confessed that he had been bribed by Mary's sister in law to smuggle rabbits and other animal parts into a hiding place in Mary's room. Still refusing to believe she had been rumbled,

Mary continued to deny it had been a hoax when interrogated by Richard Manningham and James Douglas. During these interrogations Manningham examined her and found that she still had something in her uterus, eventually, after Manningham threatened to cut her open painfully to find out what it was, Mary confessed. It turned out that it was the rest of the cat, of which parts had been delivered by John Howard in one of the first fake births. Immediately following her initial miscarriage Mary Toft had inserted animal parts into her uterus after being told how to do this, she claimed, by a gypsy who said that if she were to do that she would become famous and want for nothing. After her uterus had contracted she had started pushing meat and body parts into her vagina and birthing them by opening her legs, after a bit of theatrical moaning.

As all this was happening St Andre was busy publishing a 40 page book on Mary's strange sooterkin,[1] which was published on the 3rd of December 1726, leading to a humiliating climb down by him on the 9th of December when the hoax finally came to light. St Andre lost his position as the Kings surgeon as a result of the affair. Two years later in the houses of parliament he was to encounter Samuel Molyneux once more, and received widespread condemnation when Molyneux died in his arms after suffering an apparent fit in the chamber and it was alleged that St Andre had poisoned him and used his position as a surgeon to prevent him from getting help before he died. St Andre then eloped with Molyneux's rich widow, so there may well have been something in those accusations. Indeed, the whole Toft affair could well have been used by St Andre as cover for an affair with Molyneux's wife.

Mary Toft and John Howard appeared in court the following January. Howard was fined £800 for his part in the scam and Mary was briefly incarcerated. Ill health was to secure her release and she went back to her husband in Godalming where she later had a normal, healthy and human daughter.

---

[1] A sooterkin is a small creature that women were fabled to be able to give birth to.

THREE NOTES ON THE BUNYIP

Bob Skinner passed the following article to me:

Hereford Times December 13th 1845 page 1

VERY LIKE A WHALE. An Australian paper, the Squatters'
Advocate, has the following, under the heading, "Wonderful discovery of a new animal." - In our last number, we gave an account of the finding of the fragment of the knee joint of some
gigantic animal, which, from there being no such animal hitherto
known to exist in Australia, we supposed to be the fossil remains
of some early period.

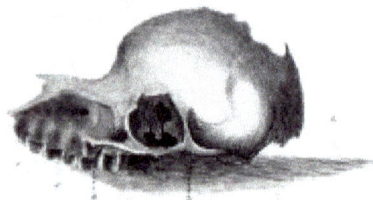

Fig 1 Bunyip skull.

The Tasmanian Journal of Natural Science 1847

Wikepedia Creative Commons

Subsequent information, however, coupled with the fact that
the bone was in good preservation, and had altogether a recent appearance, has induced a to alter our opinion. On the
bone being shown to an intelligent black, he at once recognised it as belonging to the `bunyip`, which he declared he
had seen. On being requested to make a drawing of it, he did   54

so without hesitation. The bone and the picture were then shown separately to different blacks, who had no opportunity of communicating with each other, and they one and all recognised the bone and picture as belonging to the bunyip, repeating the name without variation. One declared that he knew where the whole of the bones of one animal was to be found;another stated that his mother was killed by one of them at the Barwon lakes, within a few miles of Geelong, and that another woman was killed on the very spot where the punt crosses the Barwon, at South Geelong. The most direct evidence of all was that of Mumbowran , who showed several deep wounds on his breast, made by the claws of the animal. Another statement was made, that a mare, the property of Mr. Furlong,was, about six years ago, seized by one of these animals on the bank of "the Little River", and only escaped with a broken leg. - They say that the reason why no white man has ever seen it, is because it is amphibious, and does not come on land except on extremely hot days, when it basks on the bank; but on the slightest noise or whisper it rolls gently over into the water, scarcely creating a ripple. We have adduced these authorities before giving a description of the animal, lest from its strange,grotesque, and nondescript character, the reader should have at once set down the whole as fiction. The bunyip,then,is represented as uniting the characteristics of a bird or alligator; it has the head resembling an emu, with a long bill, at the extremity of which is a transverse projection on each side, with serrated edges, like the bone of a stingray. Its body and legs partake of the nature of the alligator. The hind legs are remarkably thick and strong, and the forelegs are much longer, but still of great-strength. The extremities are furnished with long claws, but the blacks say its usual method of killing prey is hugging it to death. When in the water it swims like a frog, and when on shore it walks on its hind legs with its head erect, in which position it measures 12 or 13 feet in height. Its breast is said to be covered with different coloured feathers; but the probability is that the blacks have not had a sufficiently near view to ascertain whether this appearance might not arise from hair or scales. They describe it as laying eggs of double the size of the emu's egg, of a pale blue colour; those eggs they frequently meet with, but as they are 'no good for eating' the black boys set them up for a mark, and throw stones at them. (1)

GREAT LAKE "BUNYIP"

What Mr Parker Saw

Rare Beetles Found

The Mercury (Hobart, Tas. : 1860 - 1954), Monday 4 March 1935, page 8. This is an abbreviated account leaving out the references to beetles.

"The bunyip" at the Great Lake has had many descriptions fastened on to her, but so far I have not heard anyone who has described it as a beetle" said Mr Critchley Parker on arrival in Launceston on Saturday morning. He was referring to the fact that he had been fortunate enough to find two rare beetles in the Great Lake, which proved that there were "other things in the lake other than 'bunyips.'

I read with interest of the reappearance of the common (female) seal, which has become known to the newspaper readers throughout Tasmania as the 'bunyip'. My original statement was made to 'The Mercury' in a short article, and was followed up by a personal interview with the late Mr Simmonds. I was crossing the dam with with the superintendent of the Government Printing Office (he being some 20 yards behind). At the fourth arch from the north end of the dam at the lake, and in eight feet of perfectly clear water, I saw a seal floating on top of the water, not more than 15 feet below me. I saw the soft beautiful brown eyes of the seal, and the yellowish fur, I marked the spot on the cement and later confirmed the depth of the water from a boatIt was a perfectly calm day. The sky was clear, and there was a bright sun. The time was approximately between 4.30 and 5 o'clock in the afternoon.

56

For 37 years I have known of the existence of a seal or seals in either the lagoon before the dam was built, and the possibility of a seal in the Great Lake itself after the dam was completed. When one is on the shore one is raised very little above the surface of the water. I have seen musk ducks and very large species of platypus, but I felt it quite unsafe to venture an opinion until I saw my seal in the circumstances related. (2)

I found the following story via Trove, from **The Sydney Morning Herald** of September 9th 1949 page 4.

Weird Animal "Uses Ears As Paddles"

MELBOURNE. - Kyneton has joined in the open season for bunyips.

Mr and Mrs L.Keegan reported to-day that several times in the past fortnight they had been astounded by what they described as an animal at least four feet long, with long shaggy ears , in the new Laurisior reserviour, adjoining their property.

They said it used its ears to propel itself through the water "at tremendous speed."

" It dives and has remained under water for a considerable distance before surfacing", they said.

" When it submerges the noise can be heard from about 20 yards away"

The Keegans are unable to say if it has fur or feathers, because they have not been able to get a close-up view of it. But they are certain of two things - it is larger than a swan and has long ears.

Mr J.Beare, a school teacher on holidays, saw the animal twice and has confirmed the description given by the Keegans. (3)

REFERENCES

1. Hereford Times December 13th 1845 p. 1

2. The Mercury March 4th 1935 p. 8

3. The Sydney Morning Herald September 9th p.4

Fig 2   Aboriginal Myths - The Bunyip. 1890.

Mike Hardcastle/Trove

NOTES AND QUERIES

A note in **History of British Animals** by John Fleming (1828) says:

"Ray takes note of the L.*viridis* , or Green Lizard…, as inhabiting Ireland. It occurs in Guernsey; and, according to Pennant, it has been propogated in England. The upper parts of the body being a rich, variegated green, the belly whitish, and the length being from 18 inches to 30 inches, distinguish it from the L.agilis. Pennant mentions a lizard, probably of this species, " which was killed near Woscot, in the parish of Swinford, Worcestershire, "which was 2 feet 6 inches long, and 4 inches in girth. The fore-legs were placed 8 inches from the head; the hind-legs 5 inches beyond these; the legs two inches long; the feet divided into four toes, each furnished with sharp claw. Another was killed at Penbury, in the same county. Whether these are not of exotic descent, and whether the breed continues, is what we are at present uninformed of." - British Zoology iii . 22(1)

In the **Statistical Account of Little Dunkeld, Perth and Kinross,** (1793) by the Rev John Robertson, Vol. Vi p. 361 is the following note. " A quadruped found in the moors at the eastern extremity of the parish, is entitled to notice as a remarkable variety of the Lizard tribe. It is about 9 inches long, the body, or trunk, is of an unusual length in proportion to the tail, which does not taper gradually from the hind feet , as in other lizards, but becomes suddenly small, like that of a mouse. The back is full of small protuberances, and guarded with a skin almost as hard as a sea shell. The eyes large, clear, and circular, like those of an ordinary trout; the jaws more than an inch in length,and the teeth so strong as to be heard making a ringing noise upon the iron point of a pole at a distance of more than ten feet. It is believed in that part of the country, that, about 50 years ago, the bite of this animal proved fatal to a child two years old. It is never seen but upon very dry ground. When irritated it expresses rage by the reddening and glistening of its eyes." (2)

ATLANTIS

UNIVERSAL DELUGE

Scientific Confirmation Of

Actual Occurrence

ATLANTIS ALSO EXISTED

The following appeared in the South China Morning Post of February 7th 1940.

The scientific confirmation that the Universal Deluge actually occurred in the days of Noah and that the mythical island of Atlantis existed 10,000 B.C is claimed to have been reached by the famed Italian scientist and astronomer, Professor Raffaele Bendandi. [1]

In an exclusive interview by telephone from his hometown of Faenza, Professor Bendandi stated to the *United Press* that his recent discovery of four planets beyond Neptune gave him the first indication on which he has been working ever since.

The youthful Professor stated : "Owing to my recent discovery I have been able to ascertain that those celestial bodies regulate our entire solar system. It is their huge power of attraction on the world axis which caused all geological movements and earthquakes. By means of careful research work on the movement of these planets I have now reached the conclusion, supported by scientific proof, that the Universal Deluge actually occurred in the year 2687 B.C as stated in the Bible. Continuing my research I have also obtained scientific proof that the mythical Island Atlantis actually existed West of the Pillars of Hercules and was submerged beneath the Atlantic Ocean in the Autumn

---

[1] "Professor Bendandi (1893-1979) was a pseudo scientist who specialised in the prediction of earthquakes." Wikipedia.

of the year 10,008 B.C.

Professor Bendandi added that that both the existence and the
date of submersion of Atlantis are confirmed by the Greek phi-
losopher Plato in his writings. (3)

Fig 1   Map of Atlantis by Athanasius Kircher 1669

Wikipedia Creative Commons

THE JAPANESE MYSTERY TORTOISE THAT WASN'T

 In September 2012 my Mum and I visited Whitby, Yorkshire and its main Museum. In the Asian ethnographical
cabinet I noticed a Japanese natural history book (see front
cover of this issue of Flying Snake.)

The book had a small plaque next to it which said: **Book of Natural History** from Japan. Dated Spring in the First Year of Genji (1864) Gift of H.P. Kendall in 1955 WHITM : ETH 63. I posted a note about this on Cryptozoology Online on October 5th asking for information about this brightly coloured tortoise as I thought it might be of cryptozoological interest but I received no replies. I asked Allen Salzberg of Herpetology Digest if he could help me identify it and with his help I found out the following from Peter Paul van dyke head of Turtle Conservation at Conservation International:

" I would not think this is identifiable - the scute pattern is unlike any real turtle (last vertebral projecting over the tail, instead of having supracaudals), the head pattern matches no current or historic Japanese species (including Taiwan, Korea etc.), and while there is a faint resemblance to *Geomyda japonica* from the Ryu Kyu islands , I would treat this as a likely composite of one or more real animal parts (a partial shell or so) sketched at a different time that this actual illustration was made, with liberal interpretation by the artists to fill in missing bits; or the artist copying from other drawings to try and produce a different posture. This is in some contrast to many of the classic woodcut prints of turtles, which tend to be clearly drawn from live specimens of *Mauremys japonica* ." (4)

According to Peter Pritchard:

"Richard: the Japanese turtle illustration is almost certainly based on Geoemyda japonica, a very prized species now protected and considered to be a national treasure. It occurs only on islands off the south end of Japan (Okinawajima, Kumejima, and Tokashikijima. The artwork is very attractive, but is only a very rough representation of the species." (5)

FLYING ADDERS

This extract is from Aubyn Bernard Rochfort Trevor-Battye`s **Pictures in Prose of Nature,Wild sport,and Humble life** (1894). It relates to somewhere in Britain, I'm not sure where, possibly Kent:

"The country folk in certain parts are firmly persuaded that the adder as it grows old develops a pair of wings and flies about. They quite believe that they have seen it flying. This diabolical accomplishment intensifies the terrors of the " death-arder." Everything that creeps and looks like a snake is a death adder. The idea arose in the church, by mistake for "deaf adder," long years before the School Board came." (6)

Now another one of those weird Chinese creatures I love:

Freak Bird

HAS HEAD LIKE A MONKEY

(Hong Kong - By Air Mail)

"With the head of a monkey, a strange bird is reported to have been caught in Hankow.

Measuring three feet in height and eight feet in breadth, the creature is said to have " fur-like black feathers, four toes on each foot, and a mouth shaped like the bill of an eagle." This was reported in the Mirror newspaper of Perth Australia on July 17th 1937." (7)

The following strange story of a Two-Legged Snake from Japan appeared in The Straits Times on August 1st 1922:

The Straits Times August 1st 1922

THE TWO-LEGGED SNAKE

Wonderful Reptile Found in Japan

"While naturalists are discussing the origin of China`s dragon, the capture of something of the sort at the summer resort of Karuizawa, Japan, may assist them. The creature is said to have been a two-legged, horned snake, and referring to it he Japan Advertiser says:-
"From description it sounds as if it was a cross between an ordinary snake and a Rocky Mountain billygoat. It is 5 feet 9 inches long and at the largest 5 inches in circumference. It has two legs about 18 inches from the tip of the tail. Nor is that all. A horn protrudes above each eye. As children were passing through the grass and bamboo at Pulpit Rock on Monday the snake stood erect and darted its fangs at them. Men came to the rescue, killed it and have spent most of the time since repenting for not capturing such a rare specimen alive. The reptile is now preserved in alcohol at a shop on the "Machi" and the people here say that nothing like it has ever been seen before in Japan. From the description your correspondent is inclined to believe that no such creature has ever been seen elsewhere. Persons who read this may be inclined to believe that it has never been seen at Karuizawa. This story, however, comes from an eye-witness,who, by the way, is a good American prohibitionist"

Writing on the subject of this discovery, an American correspondent of the N.C.Daily News[1] says - " I have seen a good many curious things in the course of my rambles but a snake which stands upon its hind legs has not so far been numbered among them , and the first impression was that it must have been some fancy born of the 35th drink or thereabouts. On glancing further down your article, however, I see that the witness was a prohibitionist, and it seems to me that if this is the effect that prohibition is going to have on people,well,

---

[1] North China Daily News

the sooner the 18th Amendment [2]is put into the discard the
better."(8)

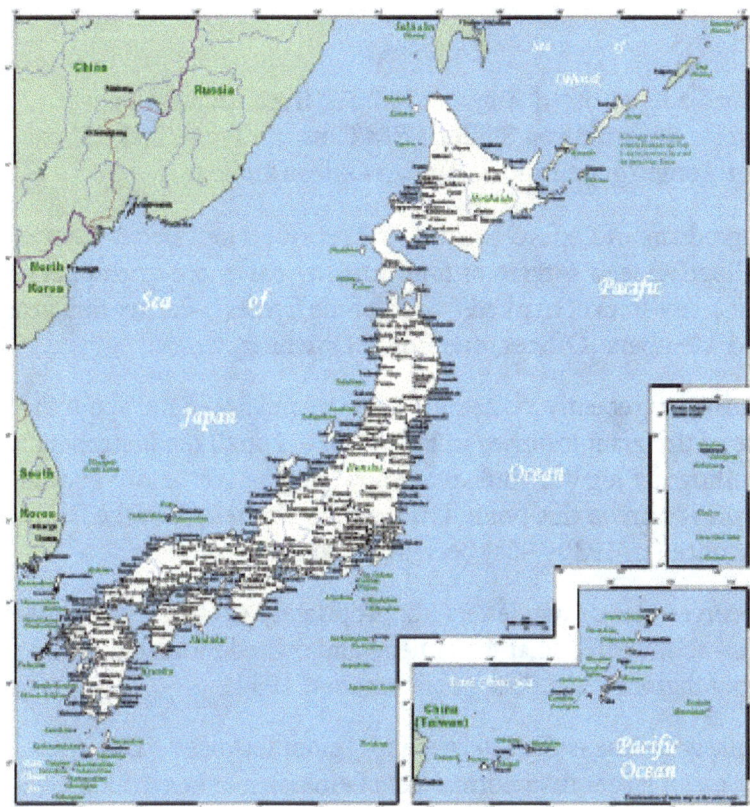

Fig 2   Map of Japan

Wikipedia Creative Commons

---

[2] The 18th Amendment established prohibition of the consumption
of alcohol in the United States of America with the law taking effect
on January 17th 1920. It was repealed in its entirety by the 21st Am—
mendment on December 5th 1933. (Wikipedia.)

THE OXFORD COLLEGES ANCIENT TORTOISE

I received a letter dated August 31st 2012 from Richard George, who wrote the article on Steller's Sea Cow in Flying Snake 3, concerning some "giant tortoises" at an Oxford College:

"As a graduate of Oxford University I receive a newsletter from my old college, Queen's, three or four times a year, more often than not rattling a beg-bucket marked " Sign of the Times". Earlier this year the Old Members 'Officer, one Emily Downing, wrote:

"A gentleman recently visited the College and asked me about the demise of the giant tortoises in Front Quad. I shall readily accept he was pulling my leg but you never know, and he left before I could really press him on this point. Can anyone else testify to the College having once (c. 1930s-40s) been home to some tortoises?"

She received one or two letters - appropriately enough from centenarians - to the effect that the college had owned tortoises of some kind, but that they had disappeared around 1950.

What intrigues me is the reference to " giant tortoises". Does this simply mean larger than normal pet chelonians, or something Galapagosesque? (9)

My friend Rob Wilkes did a bit of his own digging into this case and, working at the Bodleian Library, found out the story reproduced from *Floreat Domus* Issue 13, 2007 below:

"With the sad and unexplained disappearance in 2003 of Rosa Luxembourg, the much-loved College tortoise who had been here since the 1960s, it is high time that we welcomed a successor to her. The new tortoise sensibly waited until after the Balliol Ball to make its formal appearance, and is a surprisingly lively and engaging creature.

Whilst its existence may not prove conclusively Balliol's green credentials, or do anything for the College's carbon footprint, the tortoise is a welcome addition to the membership of this institution.

So far, 'it' does not have a name, although we are open to suggestions from Old Members and anyone else who may have an idle moment. We think that it may be a male, but – as was the case even with Rosa Luxembourg – we cannot unfortunately be sure."(10)

The book Tortoise by Peter Young (Reaktion Books) contains the following:

"Oriel College, Oxford, adopted its original tortoise as a mas-cot in about 1896, the creature becoming so familiar that it was elected an honorary vice-president of a College society. An undergraduate found a companion for it in his bed. When the mascot died in 1923 it was stuffed. So lifelike was the result that one of the fellows, finding it in the common room, took it out into the sun. The living tradition continued with two tortoises on whose shells were blazoned the College arms. On 28 May a tiny tortoise appeared in a quadrangle by the side of the other two, with Ichabod inscribed across its shell. His birth was announced on 31 May in The Times: 'testudo, to Georgina, wife of O.C.Testudo, a son (Whalley George).' He was believed to be the only tortoise to have his birth announced in The Times. George Whalley was the honorary secretary of Oriel College Boat Club, which had the tortoise as its emblem and which in Eights Week had failed to come head of the river. On the first night that the college boat made a bump (touching the boat in front of it), Georgina laid an egg. The next day, when the boat made another bump,a second egg was laid, but there were no more bumps or eggs provoking the provost's observation that we will never know whether Georgina's productivity would have continued in arithmetic or geometric progression. Only the provost's wife or daughter could properly give birth within the College precincts, and sometime after the appearance of the birth notice in The Times an academic colleague of the provost said that he was unaware that the provost's daughter had married an Italian.

Fig 1   Queen`s College Oxford

Wikipedia Creative Commons

Fig 2   Galapagos tortoise. Wikipedia Creative Commons

Testudo was much kidnapped by under-graduates from other colleges. To his displeasure, other tortoises, some having the names of fellows painted on their backs, were added to the College collection. When he died in 1949 the provost penned a short elegy:

All his slow life he kept his secret well
Of what he loved and hated and believed;
It died with him, and we whom he deceived
Interrogate in vain his empty shell"(11)

On October 5th 2012 Rob Wilkes told me: "By the way, a colleague remembers that New College had a giant tortoise in the mid-'70s that was supposed to be 100 years old." Richard George also mentioned the Oxford College tortoises to Karl Shuker who had his own "take" on the subject in the November 2012 Fortean Times:

"In case you're wondering what this has to do with cryptozoology, there is a bona fide link. Living giant tortoises are known today only from the Galapagos Islands off Ecuador and from the Indian Ocean island of Aldabra. Just a few centuries ago,however, additional species also existed on the Seychelles, but these were believed to have been wiped out - until, in the past two decades, certain captive specimens hitherto believed to belong to the Aldabran species were unmasked by DNA and other analyses to be surviving individuals of two of the supposedly lost species from the Seychelles. Consequently, if giant tortoises did once exist at Oxford University, it is just possible that they too belong to these formerly `hidden` species of `hidden` species of Seychelles giant tortoise, in turn meaning that if their bodies have been preserved and can be located, they will be of great scientific significance." (12)

A MEWING MALAYSIAN SNAKE

The China Mail of October 18th 1935 carried the following:

MALAYAN SNAKE SAID TO MEW LIKE A CAT

"An 8 foot opistoglyph, a canibalistically inclined snake, believed capable of meow-ing like a cat, has arrived at the West New Brighton Zoo. It was sent to Mr Caril Stryker, director of the zoo by Mr A St. Alban Smith, British rubber planter in the Malay States. Mr St Alban Smith wrote that the natives believed the snake was able to make feline sounds. Reptile experts are doubtful- Reuter." (13)

1. J. Fleming History of British Animals (1828)

2. Rev J. Robertson Statistical Account of Little Dunkeld , Perth and Kinross (1793) Vol VI p.361

3. South China Morning Post.   February 7th 1940

4. E-mail from A.Salzberg to R.Muirhead October 4th 2012

5. Ibid

6. A.Trevor-Battye Pictures in Prose of Nature, Wild sport and Humble Life (1894) p.239

7. Mirror   July 17th 1937

8. The Straits Times   August 1st 1922

9. Letter from Richard George to Richard Muirhead August 31st 2012

10. Floreat Domus Issue 13 2007

11. P. Young Tortoise (2004) pp 126-127

12. Karl Shuker. Alien Zoo. Fortean Times November 2012 p. 25

13. The China Mail October 18th 1935.

BOOK REVIEW

Sea Serpent Carcasses. Scotland: From the Stronsa Monster to Loch Ness. Glen Vaudrey. Bideford England : CFZ Press 2012 ISBN 978-1-905723-93-5

This excellent book is a "grand tour" around the coast and many islands of Scotland (which is dear to my heart,look at my sur name - Richard.) and its surprisingly frequent sea-monster strandings, starting with the Alba case of 906 A.D. recorded in the Irish Annals of Innisfallen, then the famous Stronsa Monster of 1808 with a number of intriguing illustrations and quotations from the time of the incident. The Loch Ness Monster, not surprisingly, makes an appearance, but not the famous cases from the start of the "modern period" i.e. 1933, but 1868, a hoax reported in the Inverness Courier - the corpse of a northern bottlenose whale. The most recent case is that of a strange lump of flesh washed up at the Bridge of Don in 2011. The book includes Glen`s unique plastacine (?)carcass creations and other art-work,also there are copious maps and a Conclusion and Random Musings,Bibliography,Index. Dr Devo says -buy it for a friend or relative for Christmas!

Letters to Flying Snake

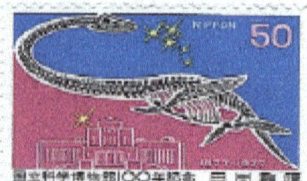

CAT CONVENTION

The following letter was e-mailed me on February 16th 2010 from Patrick Foord in response to a letter I had published in Fortean Times # 253 September 2009 concerning cat conventions:

Cat "conventions"

I am interested in reports of cat "conventions (FT # 245:75) - that is, gatherings of domestic cats in a circle, usually at night, with a larger cat sitting in the centre in an open space, apparently directing proceedings. If anyone has any further information or can point out accounts in the literature, please write in.

Richard Muirhead

Dear Mr Sutton

I was reading a back issue and chanced upon the letter from Mr Muirhead regarding `cat conventions`, page 73 of that issue.

If possible would you be kind enough to pass on my experience of this phenomenon to this gentleman.
I was aged about 14, and was awakened in the small hours of the morning by incessant noise from a number of cats nearby. My parent's back garden was circled by a high wall, and beyond one side was a road. The noise seemed to be coming from that direction, so I grabbed the top of the wall and hoisted my head above the brickwork and peered into the street-lit road.

I was astonished to see a large circle of cats surrounding two cats in the centre who appeared to be facing each other off. The proceedings appeared to be under the authority of a large ginger tom who was sitting in the circle to one side of the other two, and apparently overseeing the confrontation. At the appearance of my head over the wall all of the cats ran off in fright, including the apparent contestants. All, that is,save the large ginger tom who still sat,looked at me for a few seconds, then slowly got up,and very disdainfully turned his head away,and walked off slowly down the road with his tail erect. I have never since read of any similar event, but clearly since Mr Muirhead expressed his interest est it is not such a totally unknown phenomenon as I had supposed

Kind regards   Patrick Foord

DR WHO AND FLYING SNAKES

A letter from Oll Lewis dated September 30th 2012

As readers of "Flying Snake" will know two well known sightings of snakes with wings, or Gwybers, have occurred in South Wales. Interestingly the TV show Doctor Who has filmed where these sightings were said to have happened since it was re-launched several years ago. Until recently scenes like the Tardis interior shots were filmed in the Upper Boat Studio near Culverhouse Cross and Exterior and on location shots for the episode "Tooth and Claw" were filmed in Penllyne and Penllyne Castle. By coincidence the episode filmed in Penllyne featured a cryptid too in this case a Werewolf. Sad to say it wasn't the best of episodes, but if you're interested in seeing the locations of the Penllyne Gwiber sightings and can't go there in person then the episode is worth watching.

Oll Lewis

BUTTERFLY AND MOTH PHOBIA

Richard George wrote to me on May 11 2012 ;

Dear Richard,

RE *Notes and Queries* in *FS* 2...

I've never heard of lepidopterophobia with butterflies of British dimensions, although I can certainly imagine people being afraid of tropical ones.

Moths, on the other hand...both my mother and I are distinctly uneasy with them. In Mum's case, it stems from a childhood encounter with the rare Death's Head Hawkmoth, *Acherontia atropos,* in an overgrown garden in Birkenhead. I had a similar not-quite-panic attack a few years ago in my home when confronted with an Old Lady Moth, *Mormo maura*. It appeared from nowhere on my landing and seemed to chase me around.

And yet this moth (*Mormo maura*) was little different in size to a butterfly. Do we tend to fear moths more because of their association with night? Or is it, to paraphrase George Harrison, something in the way they move?

Now I'm off and running on music, one of the great lines of all time is Jim Morrison's "The scream of butterfly..." from *When The Music's Over*. That *is* scary.

With best wishes,

Richard George

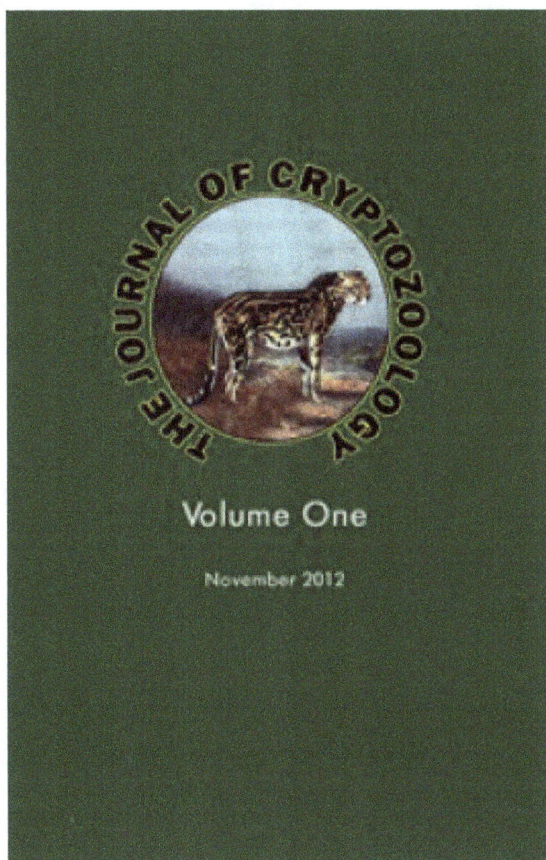

Volume One

November 2012

http://www.journalofcryptozoology.com/

At £6.99 plus postage

Not for sale from Flying Snake

Flying Snake

A Journal of
Cryptozoology, Folklore and Forteana

Volume 2   Issue 2   July 2013   £3.99

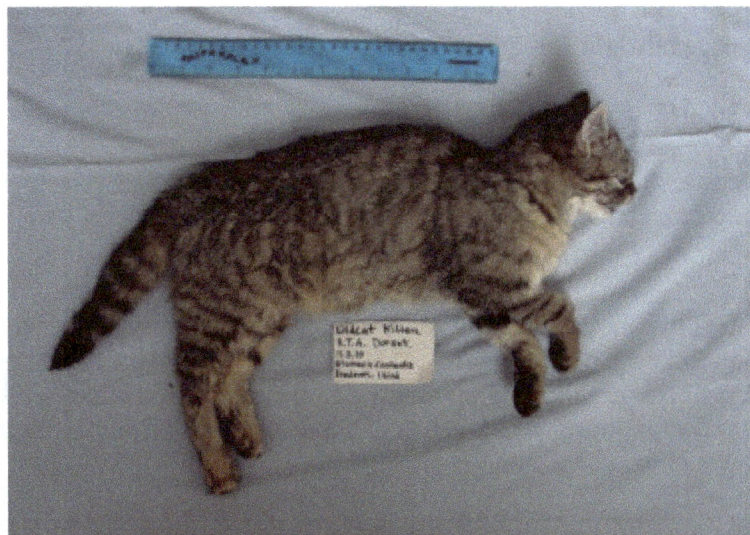

A Dorset Wild Cat? • Unidentified Submarine Objects •
Cumbrian Parrots • Hong Kong Tiger 1940 • Locusts in
Israel•Horned Rats• Chinese Alligator • And more!

ABOUT FLYING SNAKE

Flying Snake is available from:
Richard Muirhead
Flying Snake Press,
112 High St,
Macclesfield,
Cheshire,
SK11 7QQ
UK
http://homepage.ntlworld.com/richmuirhead/cryptozoology/

Tel: 01625 869048

 Mike Hardcastle ,Sub-Editor,NSW Australia .   Zvi Ron Israel correspondent.
www. steampunknaturalist. com   Carl Marshall   Zoological Consultant.

Please feel free to contact me if you want to reproduce anything I have written. If
you want to reproduce other authors' works, I will try and contact them on your
behalf and get back to you. The opinions of authors other than myself do not
necessarily reflect my own. Blank authorship indicates essay from a collection of
newspapers,etc,or unknown author(s).

PAYMENT

Subscriptions: £3.99 per issue, £12 per annum. Full colour PDF via e-mail.
£3.99

Payment for however many issues you (and your friendly neighbourhood flying
snake) would like to purchase can be made by means of PayPal on my web site
(see url above) Checks and postal orders from within the U.K. should be made out
to Richard Muirhead NOT Flying Snake. Checks will not be accepted from abroad

CREDIT

The front cover images were provided by Jonathan McGowan. Thanks.

CONTENTS

DR DEVOS DIARY

*"For I pray God for the introduction of new creatures into this island.
For I pray God for the ostriches of Salisbury Plain, the beavers of the
Medway and silver fish of Thames." Christopher Smart `Rejoice in the
Lamb`.Poet,naturalist,lunatic (1722-1771)*

Well hello again from the increasingly sub-tropical ambience of
Macclesfield (I provide you with the surface synoptic chart for 00 UTC
today, 12-7-13 just to prove it!) where wilder-beast frolic in the mid-
Summer haze and a dozen hummingbirds flit to and fro besides the dieing
buddleia I have neglected to look after about 30 yards beyond this office
window.

First of all I am very sorry for the delay in publishing Flying Snake 5,this
is because of computer problems,general mayhem and disorganised living
(e.g coming home at about 3a.m after clubbing which has taken its toll on
my 46 year old mind!) however I hope you enjoy this smorgasbord of
cryptozoological goodies. I would like to thank Carl Marshall, Mike
Hardcastle,Bob Skinner, Jonathan McGowan,Nick Redfern, and Zvi Ron
and the various anonymous newspaper journalists with the long defunct
and presumably fossilized North China Daily Herald , Chinese and

Western eyewitnesses in late 19th Century Shanghai and the alligators for
turning up for their major part in this issue. Flying Snake 6 onwards will
be different, A5 size still,30 pages and 6 x a year the same content but

PLEASE can I have some female contributors? On June 24th Alan
Myers,Devo`s original drummer died,so is this the Beginning of the End
of Everything Now or will the Ape Still keep on Regarding it`s Tail? Who
knows? I wish Animals and Men a successful re-launch with issue 50 and
everyone else known and unknown to me working on cryptozoological
projects at present. 2014 should see the publication of Lizzy Clancy and
my `Mystery Animals of Old Lancashire` (provincial title) and my
Muirhead`s Mysteries blogs in book form. I leave you with that.

4

A Wild Cat in Dorset 2010

JONATHAN MCGOWAN, MAX BLAKE, RICHARD MUIRHEAD

Whilst at the Weird Weekend in Devon in 2012 I got talking to Jonathan McGowan, a naturalist and taxidermist from Dorset. I was fascinated when he told me he had found a road-kill wild cat kitten near Martin Down in Dorset, which itself is near Cranborne Chase wood because I had visited the former and gone to school in the latter place as a child between 1975 and 1980. This article is about a debate in late 2012 between myself, Jonathan and Max as to the exact identity of this mystery feline. The following pages contain Jonathan's opinions on the wild cat in Dorset and the whole debate.

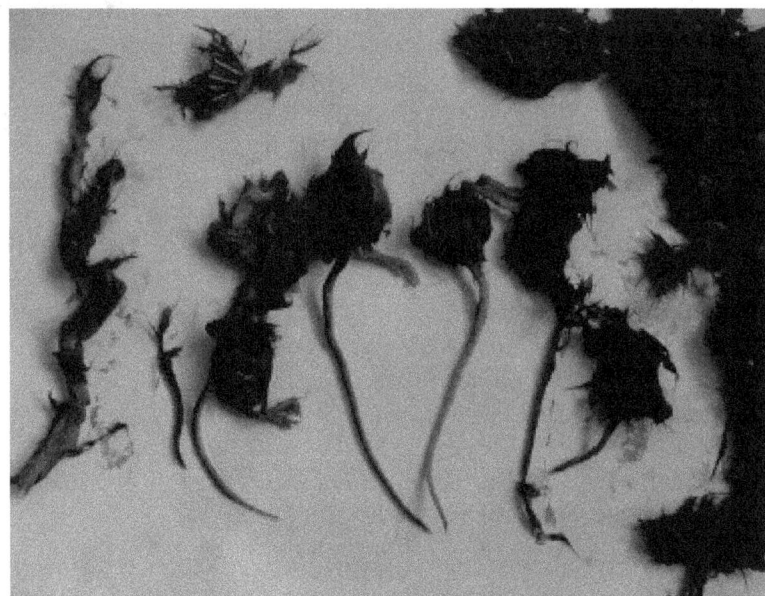

Fig 1 Stomach contents of the Dorset "Wild Cat" © Jonathan McGowan

The English wildcat.

Felis sylvestris.

In southern England.

It is not known as to when the species became extinct in southern England. It was one of those assumptions that slowly faded into view over a along time period. Not much was done about it as the wildcat, like other predators at the time were classed as pests and had no conservation status. It was thought that the wildcat disappeared long before the Scottish animals became a conservation talking point. All emphasis was then put on these animals as it was thought that Scotland was the only place where wildcats lived. This was the nineteen sixties. It was later thought that wildcats in southern England had been wiped out before the turn of the century. There was no data on this species here and although there were many records from gamekeepers and landowners during the nineteen twenties, they were overlooked or thought as mistaken identity. The unofficial records came from the hunting and shooting fraternity who would often see or shoot an animal,several reports of wildcats were from rabbiters with ferrets that would bolt the animals into nets,two reports came from the Cranborne Chase area of Dorset during the late nineteen twenties. Several such reports can be seen in various country books and novels such as old game keepers tales and similar books. The hunters would describe a large grey brown tabby looking cat with vertical thin stripes, with a bushy tail that was so ferocious that they could not cut the nets to release them, often just shooting them as they were so tangled up. Wild cats often den up in old fox earths or unused parts of badger setts and it is well known to rabbiters that rabbits often dwell in these places so netting any underground earth that has not got obvious signs of badgers or fox living in it is worth netting and putting down ferrets to flush out possible rabbits. Wildcats were keeping a very low profile, they were being targeted from all sides. They became even more elusive but living in the remote farming areas especially in the hilly regions with lots of thick woodland. So who told everyone that they had become extinct ?

Well it was possibly because there were no real reports from the growing conservation fraternity or naturalists had not encountered them. Secondly, there were so many feral domestic cats about that the chances of them still being around were virtually impossible as genetically thoroughbred animals.

That was the assumption. Cranborne Chase is a large area of remote hilly wooded areas. There is much natural grassland and native broadleaves forest and also planted conifer woodlands. Much of the high areas are uninhabited by people except for farms which are often a long way from the land that is owned. The chase covers the mid county to the north East and so takes in a huge swathe of the county. I have lived on the edge of the chase and spent a lot of time in many of its areas as a child. It is the area where I observed pine martens when I was a teenager, another animal thought to have gone extinct a long time ago. Strangely the two animals overlap in regards to sightings by country folk during the nineteen thirties and forties. During the war, the animals were observed more frequently as game keeping was held back for a while during the war and maybe the remnant populations were able to sustain themselves long enough into later conservation years of the nineteen sixties. When I lived in the area I observed pumas hunting and breeding in the area, and I was the only person who saw these timid shy animals because I was out at night watching owls and badgers. If the local people knew nothing of these naturalized foreign cats, then what hope have they of knowing that wildcats were also present? At night in many areas of Cranborne Chase, one can walk at night through many areas without seeing a human or even a car on the night roads. Any nocturnal animal could be living there unobserved. So comes my encounter which may blend the ideas, or assumption that the wild cat did not perish but still lives on , possibly diluted in terms of genetic purity.

On the morning of the 11 of March 2010 I drove out to search for road killed animals to eat and to use as taxidermy. I often travel on a productive rout which is a fifty mile round route from my base in Bournemouth. Cutting through a small portion of the southern tip of the chase often conjures up many pheasant carcasses on the road as the area is hot in regards to game rearing. There are many dead animals to be had from this area around Sixpenny Handly and Cranborne including stoats, rats, squirrels, three deer species, foxes, badgers rabbits and many hares. Many raptors also get hit on roads such as owls and buzzards. It is also a good area for ravens. So it is an area where many predators still thrive despite the game rearing. I saw a shape in the road outside a large track of forest and fields behind the National nature reserve of Martin Down, an area of wild natural grassland with much woodland and high biodiversity. I pulled up to the dead animal wondering what it was, and when just a few yards from it I realised that it was a kitten but on picking it up I at once realized that it was no normal tabby cat but had all the hallmarks of being a wildcat ! I bagged it and at home I looked closely at it. It was certainly a kitten, it had a kitten face and was still a bit short in the leg, but as legs go, they were longer than a normal domestic act. The tail was short and slightly tapered just like a wildcat kit. The baring was low down with a gap without bars and the dorsal line was broken up and ended at the tail base. The body consisted of vertical slightly waving thin bands on a grey background. The under fur had a sandy tint to it. The feet were black and it had broad inner leg stripes. The ears had a lighter and dark area more obvious than most domestic cats. Its jaw was broken and what was obvious was the fact that the teeth were still baby teeth, yet the animal was the size of a small female tabby, which meant that when she would have been fully grown, she would have been larger than an average tabby cat. Her fur was very thick, not long in the same way as a long haired cat, but dense, thick hair that I have never found on domestic, not even feral cats. Her body was very chunky and strong. She was broad, unlike domestic cats.

 If she had been found in the Scottish highlands, then nobody would have disputed what species it was, in fact I have seen many photographs of alleged wildcats that look more like tabby cats than this specimen that I found. In her stomach were the remains of wood mice, field voles, a rat and a thrush. The remains were parts and not the whole animals suggesting that she had been with either a mother or siblings when eating. She would have been the product of a late litter possibly due to the death of the mother's first litter. I preserved the kitten and still have it now at home with me. I await somebody to do a genetic analysis on her.I have no doubt that wildcats still exist in southern England. I have seen photos of other road killed animals from Devon , Dartmoor to be exact. They look slightly different from my Dorset specimen but it is clear that they have some wildcat in them. Maybe the pure animal disappeared a long time ago but its genes still survive within the most remote areas of southern England, this is the same from Scotland. It is thought that no pure wildcats exist any more, but still the conservation status is hot for the animal. If so, then why has nobody got onto the case of the wildcat in southern England ? surely it is even more deserving of protection. I believe that a lot of it comes from the name 'Scottish wildcat.' It denotes exactly where the animal is found blocking out any other possibilities. Where the animal is found in highland regions of Scotland, it actually prefers the farming areas of mixed fields and woodlands to rugged mountains; Doesn't this suggest something ?

So now the debate:

On November 16th from Mike to me:

Hi.I am a bit annoyed about peoples ideas without knowing all about the cat. the tail is not very pointed and in some of the pics the fur was wet making it seem more pointed, besides, those in the know about wildcats should know that they actually have pointed tail tips until they are at a certain age. those who have read mike Tomkies books will know that and also comparing the pics he took of genuine wildcat kittens and their tails would suggest that they are also pointed until about eight weeks old when they start to bush out. my kitten was about eight weeks old I guess.

On December 3rd 2012 I received the following e-mail from Max Blake:

" Hi Richard,
I don't know much about wild cats, but I do know some stuff about their genetics.
Wild cats can interbreed with domestic cats, so much so that there are hardly any pure bred wild cats living in the UK. The 'best case' scenarios put the number of pure bred cats in the low hundreds, but other studies have suggested that this is too optimistic and there are actually far fewer than this, possibly lower than 50. Because of this, the majority of wild cats will show some characteristics of domestic cats. As well as this, the opposite is true, so there are domestic cats that look very similar to wild cats.
There are no true wild cats down near Cranborne Chase, so I would hypothesize immediately that it is a domestic cat. Though the body markings are very similar to those on a wild cat, these are highly variable anyway (see the video below). The best visible characteristic for identifying the cat is the tail. It has well defined black bands, and a black tip, a classic wild cat colouration. However, the tail is completely the wrong shape: it is fairly short, thin, and it ends in a taper to a point, completely unlike a true wild cat.
http://scottishwildcats.co.uk/identify.html
Another problem is that there is only one photo, ideally we need a photo of the dorsal side of the animal to analyse. Because of the above, I would encourage you to assume that this is a domestic cat until it is proven other wise because this is the simplest solution. Because the tail

is very wrong for a wild cat, I would say that this is a domestic cat.
What are your plans for the article?
Max.

>From: richmuirhead@ntlworld.com
Date: 03/12/2012 17:51

To: <jrmczoo@tiscali.co.uk>
Subj: CRANBORNE CHASE 2010 WILD CAT

Hi
Jonathan

Just been thinking of that 2010 wild cat kitten, the
photos of which you
sent me. Did you ever have the hairs tested for
DNA?

Jonathan replied: No I did not I was hoping that the Scottish wildcat group
may have been able to do it bit as it is not from Scotland I don`t think they
were particularly interested but maybe Someone will ! On December 4th
the correspondence continued, from Max:

Hi Richard,

Yes I did, I was too busy to reply. As we explained in the talk, we did get
the lynx DNA tested, though as we explained there was contamination
present as the pelt was extremely old, and it was essentially useless. I am
still in contact with the researchers, it cost us nothing as we were working
on a paper for it. If you found someone with a database of cat genetic
sequences, it would be really easy for them to test it. However, given the
virtual impossibility of it being a wild cat, I would guess that they
wouldn't be interested (though I would love to be wrong!).

I really think we need more photos of the specimen from various angles to
rule out it being at best a hybrid. Based on the one photo, it is still my
opinion that the specimen is a normal cat, and is not a hybrid of any sort.
We have to rule out this possibility before looking at DNA to give us an
identification.

10

On December 5th 2012 Jonathan sent me this e-mail:

Hi Richard

I am somewhat disappointed with Max`s comments and I do not think
it is right to just say it looks like a normal cat. And to think that I
would just pick up a dead cat and think that it is a wildcat is equally
absurd ! in my career as a naturalist and taxidermist, I have both found
and skinned many domestic cats. I pass many dead cats killed on
country roads, obviously ferals and think nothing of it. The reason as
to why I even stopped at this one was because it was different, not
looking like a normal cat. that was it ! and on taking it and looking
closely I noticed several interesting points. It was a kitten, albeit very
large,it had the markings of a wildcat, and yes I am quite aware that
many tabbies also have these more natural markings, but its tail was
short and thick its fur was very dense and thick,it had longer legs than
a typical tabby, its ear markings were clear, it was chunky, very broad
and muscly, its hind feet were also larger than its forefeet. it had
several head lines, an indistinct dorsal line that ended above the tail
base, basically all the typical wildcat features one could wish for. The
size alone makes it unusual! its teeth had not grown and the canine in
the bottom jaw was only partly erupted from the jaw. All of these
attributes make the animal an exception to the rule, and with the fact
that many wildcat looking cats are being found in southern England,
along with the area where I found it being the last place in southern
England where they were recorded in the past does make it a bit
interesting surely ? I have always stated that it could be a hybrid
anyhow, even though it looks more like a wildcat that many pictures
depicting typical Scottish wildcats ! It may be just an unusual feral
moggy but if it is then people must look at the Scottish wildcats in a
different light! And other so called wildcats across Europe .Here is
another pic of its back, note the chunkiness of it, this is a real feature
and is not padded out to look chunky.

Jonathan to me, again, this time on December 6th:

Hi Richard. I just sent Max an email, so its ok, he emailed me earlier.
I'm not too bothered if some folk do think that's its a normal cat anyhow. I've
had it for nearly seven years. Moths are eating it and its maybe just a genetic
throw back an old typical morph.

And then finally on December 14th:

Hi. I am a bit annoyed about peoples ideas without knowing all about the cat.
the tail is not very pointed and in some of the pics the fur was wet making it
seem more pointed, besides, those in the know about wildcats should know
that they actually have pointed tail tips until they are at a certain age. those
who have read Mike Tomkies books will know that and also comparing the
pics he took of genuine wildcat kittens and their tails would suggest that they
are also pointed until about eight weeks old when they start to bush out. my
kitten was about eight weeks old I guess.

Fig 2 The Dorset "Wild Cat" ©
Jonathan McGowan

Early Repots of the Chinese Alligator 1878-1895

Dec. 21, 1878. The N.-C. Herald and S. C. & C. Gazette.

NORTH CHINA - BRANCH, ROYAL ASIATIC SOCIETY.

A meeting of the above Society was held at its rooms in the Upper Yuen-ming-yuen road on last Friday, when notwithstanding the inclemency of the weather a fair number of members assembled to hear a paper by Mr. A. A. Fauvel on the Alligators of China. As no other business was before the meeting, the President, Mr. T. W. Kingsmill, at once called on Mr. Fauvel to begin his reading. The author, who had been at pains to illustrate this, the latest accession to our knowledge of the geology of China, exhibited in addition to the stuffed alligator which has been for some months in the possession of the Society, a live specimen obtained in last October from Chinkiang, as well as a cranium and skeleton and a stuffed crocodile for comparison. The paper commenced with a philological disseriation on the names by which the saurians of China have been known at various times, and the specimen on the table, between five and six feet in length, was identified with the t'o 鼉 or ngoh 鱷 of the old writers. Amongst mediæval writers, Marco Polo seems to have been the first to mention the Alligator in China, calling it, however, a serpent, and describing it with many fantastic accompaniments. Later writers have passed over the animal, and even Du Halde contains no allusion to it. Amongst the moderns, the late Mr. Swinhoe seems to have been the first to allude to its existence; and in 1869 a specimen was exhibited in Shanghai city,

and was described in the papers of the day as probably a new species, Père Heude, more recently, was nearly the possessor of a specimen which he only lost through his servant wrangling about a couple of hundred cash. At various times reports came down of crocodile-like animals being seen in the Yangtze, and a specimen in the Museum, really an Australian or East Indian one, was supposed to have come from the Yangtze. Fortunately, the donor was able to indicate the source from whence it came. Mr. L. E. Palm, of I.M. Customs, was, however, the first to obtain a genuine specimen, which arrived in Shanghai at the beginning of last summer, and was presented to the Society. A careful examination soon showed that it was no crocodile, but a genuine alligator, a most interesting fact, as hitherto no alligator has been met with in the old world, the genus being supposed to be confined to the rivers of America.

Mr. Fauvel then explained from the specimens, and by means of careful drawings, the peculiarities of the genus. The Chinese animal seemed to resemble most the Alligator Lucius of the Mississipi, but as it seemed specifically distinct, he proposed in the meanwhile the specific title of Sinensis, until further research should establish or disprove the distinction.

The paper led to an interesting discussion, which elicited some further information as to the occurrence of alligators in China, and the meeting was closed by a vote of thanks to the author, proposed by Mr. Bailey, U.S. Consul-General, which was carried by acclamation.

THE NORTH CHINA HERALD - N.C.D.H.

December 21st 1878

REVIEW.

Journal of the North-China
Branch of Royal Asiatic Society,
New Series No. xiii.—Shanghai,
1879.

The first article in the journal is by Mr.
A. A. Fauvel, and is a permanent addition
to our knowledge of the Natural History
of China. This paper contains all that is
known on the subject of alligators in this
country, and Mr. Fauvel has the credit of
showing that these formidable animals are
to be found in the old continent. Up to
the publication of this paper, it was believed
that alligators were only met with in the
Neotropical and Southern part of the
Neartic regions, from the lower Mississippi
and Texas through all tropical America,
but not in the Antilles. Mr. Fauvel has
discovered that the alligator, or as the
Chinese call it the "earth dragon," is a
native of Asia, and has had the distinguished
good fortune to send the first *Alligator
Sinensis* to the Paris Museum. This single
discovery is enough to show that the
Natural History department of the N.-C.
B. of the Royal Asiatic Society is not inert
or useless.

N.C.D.H.   October 31st 1879

Swimmers in the Whangpu will be glad to know that the alligator has been captured. It is to be hoped that the River Police will be instructed to see that he is not put back in the river.

N.C.D.H. July 15th 1887

We are informed that an alligator, who was doubtless making his way to the river, was arrested and nearly killed by two policemen on Broadway at 5 a.m. on Monday. Life in Shanghai will become still more exciting if we are liable to be attacked by these unscrupulous monsters in our streets as well as in our river, and children will have to be sent to school with an armed escort. It is true that these alligators are not very large, but the jaws of one recently caught measured twenty-one inches - span when opened, quite enough to give a very effective bite, though it is improbable that they would attack a human being, except in self-defence.

N.C.D.H
July 22nd
1887

Some days ago Mr. Oelkers, the foreman at Tung-ka-doo Dock caught an alligator on the dock premises. On Monday he saw another, at least eight feet long, lying on the boat slip, but before he could catch it, it escaped into the water. We would therefore advise people to look out well when they go into the river to swim.

It is astonishing how quickly Mr. A. A. Fauvel's monograph on alligators in China, published in Shanghai in 1879, has been forgotten. Mr. Fauvel shews in it that alligators have been known in China from the earliest times. They are mentioned in the "Doctrine of the Mean," among the wild animals living in the waters of China ; it is stated that drums are made of their skin, and that the sound given by the drum is like the alligator's roar. They have constantly been found in the Yangtze, and the native writers were well acquainted with their habit of burying themselves in the mud in the winter and coming out in the summer. A very accurate drawing of an alligator, which is reproduced by Mr. Fauvel, used to exist on a marble tablet in *Hai Shen Miao* Temple at Silver Island, and is probably still there, and a live one used to be kept in a pond by the priests. These priests told Mr. Fauvel that alligators are often found in the Yangtze by the native fishermen, in whose nets they become entangled. The Chinese alligator, which Mr. Fauvel decided to be a separate species, and a specimen of which can be seen in the Museum, is a small reptile, averaging only five to six feet in length, and, according to our author, he "appears very slow in his movements being nearly always in a half torpid state ; in the summer time when molested he is inclined to bite, but is never first in the attack."

N.C.D.H

July 29th 1887

Our old friend the physiognomist who has his stall at the entrance to the temple to the God of War in the native city, has just published a little pamphlet, which he retails for six cash, on the subject of the appearance of alligators in the Whangpo. The pamphlet which he published some time ago on the injury which foreigners had done to the Feng-shui of Shanghai, proved that, as an old fashioned literate, he had the greatest contempt for the barbarians from beyond the four seas; and he is now delighted to recognise, in the unwonted appearance of alligators in this river, a sign that the days of foreign occupation are numbered. He has noticed for many years that the natural sphere of the foreigner is the water, and it is on the water—by supplanting the native junks by steamers—that they have done the greatest injury to the natives in Shanghai; and it is therefore proper that the signs of their approaching downfall should come from the water. The alligator is a manifestation of the water dragon; he is therefore one of the tutelary geniuses of the country, and he has displayed himself to prove that the tutelary powers of China have not deserted her. Nothing, he asserts, could have been more appropriate that their appearance this year: we have in sport connected the appearance of these monsters with the Jubilee of the Queen, and what we have done in sport, he does in earnest. In the pride of our hearts we erected an enormous tower to commemorate the completion by our Queen of fifty years of reign; we carried on our rejoicings as freely and unconcernedly as if the land here belonged to us, and we, intentionally to his mind, drew a tacit contrast between the stability of our throne, and the comparatively brief reigns of recent Chinese emperors. Now the Nemesis is upon us, and the water-dragons are sent to warn us, and comfort the hearts of the native conservatives. Everywhere he hears, and his habit of extracting information from all the visitors to his stall keeps him well informed, that even in our own base and mechanical arts his countrymen are rapidly superseding us. The land is covered with a network of telegraphs, railways are being built by the Chinese, China's armies are as powerful as those of any Western power, while her navy is already superior in numbers to any foreign squadron in the China seas, and will soon be still stronger. He is ready to allow that in the reign of Hien-fêng a lethargy had fallen on China, and she required waking up; the instrument chosen being the rough and violent barbarian. Now that she has been aroused, and has renewed her youth, and all uncertainty as to her future is dispelled, with such men as Li and Chang and Tsêng at the helm, the barbarian is no longer a necessary evil, and he is glad to be able to announce to the over-ridden and long-suffering denizens of Shanghai that the rule of the barbarians draws to a close, and that the appearance of alligators in the Whangpu, is as the writing on the wall to Belshazzar. We have merely given a rough sketch of his conclusions; it would be too tedious a task to follow his reasoning, his proofs being derived from the interpretation of the arrangement of the scales on the water-dragon's back; proofs which are incontrovertible to anyone who has made the Pa-kwa one of the studies of his life. It was certain that studious Chinese would find some mystical meaning in the sudden appearance of these reptiles in our river; our writer's explanation is a very natural one, and we hear that his pamphlet is having a large sale and that the Shanghai Tract Society is about to publish and distribute an antidote to it.

There is a wide-spread belief in the power of some animals to take other forms, as the fox can take the form of a young and beautiful girl, rats can change to bats, sparrows to tree-frogs, kites to pigeons. Fire-flies are transformed, according to popular belief, from rotten plants, and there are numberless other transformations believed in ; but hitherto, although sharks have been known to change to tigers, there had been no instance recorded of the reverse process having taken place, and a tiger turning into a shark.

Early this month a man at Ch'ao-chow Fu, a large town in Canton province, up the Han or "Swatow River," some 40 miles from Swatow, a returned emigrant from Singapore, began to miss his pigs, fowls, dogs, etc., and at first suspected that some human agency was at work in the person of a poor neighbour, until the presence of blood, fur and feathers on the ground showed that a wild beast was the robber. So he dug a pit-fall and lay in wait.

One evening at twilight, sure enough, a large animal came prowling round and tumbled into the pit, and on going out to see, he found he had made a prisoner of a tiger. He and his friends after great trouble, got it into a cage, and were intending to send it to Singapore as a present to the British Governor ; but the tiger in cage displayed such terrible fury, roaring, clawing and biting, that, fearing it would burst out, they determined to weight the cage with stones, sink it in the river, drown the tiger, and then take off its skin.

They sunk the tiger, cage and all, and the next day hauled it up, when to their amazement there was no dead tiger in the cage, but a live alligator, and the tiger's empty skin lay in one corner. They took the alligator home in the cage and killed it, and sold the skin. Have not we all met some people, fierce and cruel in life, unreformed until the hour of death, and are they likely in a future state of existence to assume a better change than this Swatow tiger ? The tale, though it may not be true, has thus its moral.

Mr. W. B. Pryer writes as follows to the North Borneo Asiatic Society:—My name having been mentioned in connection with the discovery of the only alligator existing in Asia it may be interesting to some of your readers if I describe how it occurred. I was Curator of the Shanghai Museum for some twenty months or more from 1874 to 1876 and some time during this period a reptile was brought. I turned up Gunther's *Reptiles of British India* and other authorities and somewhat reluctantly came to the conclusion that it was an alligator; that a live alligator could occur in the Yangtse seeemed impossible in the face of Gunther's want of mention of a single species in Asia anywhere. But some one suggested that it might have been brought over in a ship and then escaped or been bought by a Chinaman and freed in the same way that they free turtles, and with this explanation we had to be content; but after a lapse of some months a second one was brought. This seemed very remarkable indeed, but one swallow does not make a summer, and two alligators seemed rather slender evidence on which to flaunt the American genus *alligator* in the face of all the authorities as an inhabitant of the old world. So nothing was done, and just before I left China a third specimen was obtained, but I was too busy to give any attention to it. To the best of my recollection none of these specimens were over six feet long, and if I remember rightly the first one was brought in by some one connected with one of the River steamers. As the occurrence of crocodiles at sea has been noted in the *Field* once or twice as rather strange, it may be interesting to mention that at least one species in Malaya is a salt-water one solely and may be found anywhere where there are islands although they may be some miles from the mainland; it is much darker in colour than any fresh-water species I have seen.

An alligator was caught in the river on Monday by some of the sailors belonging to H.M.S. *Esk*. The saurian is between seven and eight feet long, and takes kindly to pork and passes a listless existence in a bath tub.

CHINKIANG.

Our little Settlement was somewhat startled on Wednesday afternoon by a report that an alligator had been caught in the Yangtsze opposite the Settlement. The occurrence was so extraordinary that at first few of us gave credence to the report, but seeing is believing ; and, thanks to the kindness of its captor, I am now able to send you the following particulars of our strange visitor.

At about 1 p.m. of the afternoon in question, one of our community, who happened to be on board the China Navigation Company's hulk Cadiz, saw a black object floating in the water, and drifting down with the tide. It was at the time supposed to be a seal, and four of the hulk coolies gave pursuit in a small sampan. On coming up with the chase, there was a great deal of "Ai-yahing," amongst the pursuers, and one man, raising a bamboo, struck a blow at the head of the beast, which, however, had no other effect than causing it to dive and disappear from view. About half an hour afterwards the animal was again seen, and, as before, floating on the water, by a lady and gentleman who were returning from a visit to Silver Island, but they passed it at some distance, and its true character was not suspected. The alligator (for such we now know it to be) appears then to have drifted down with the stream until opposite the Consular Bluff, when, presumably try-

the Consular Bluff, when, presumably trying to reach the shore, it was caught by the up current and floated up to the Hulk of the S. S. N. Co. Here Mr. R. Talbot Williams espied it, and getting together some of his boatmen, gave chase in a gig. Our friend seemed to be perfectly unaware of the approach of his enemies, and made no effort to escape until the noose of a rope was securely made fast to his tail, when he immediately dived, but being caught up with a round turn, came again to the surface, and lashed his huge tail again and again in his happily futile efforts to escape. Finding this weapon of no avail, the animal brought its voracious jaws into play, and made strenuous efforts to burst its bonds. But all was in vain; and the prize was dragged ashore in triumph. I had the pleasure of seeing the monster this morning, and an uglier brute I hope never to meet with. From the tip of its nose to the extremity of the tail it measures 6 feet 4 inches, and its weight is close on 200 lbs. It is now placed in a small artificial pond in the American Consulate, where it lies apparently contented, for it makes no effort to escape. When disturbed, by the prodding of bamboos, sticks, &c., (we do not like to venture too close) it raises its ugly head ; and the vicious look of the eye, combined with the ominous crash of the huge jaws as it brings its formidable teeth together, make one spring back with a sense of fear. Of course,

every one is anxious to see it, and I hear that it has been presented to some Chinese, who will probably make a good sum by exhibiting it.

Now as to how our stranger got here we are all conjecture, for we believe—in the absence here of any zoologist of repute—that reptiles of this species have hitherto been quite unknown in the Yangtsze waters. The most likely story I have heard concerning the beast is this. About five years ago Messrs. Canny & Co.'s Compradore purchased—from whom I am not able to ascertain—two young crocodiles (鼉魚) or alligators, which he presented as a gift to the priests on Silver Island, to be turned into the river, but with what purpose the request was made is best known to the donor. These beasts then measured respectively 3 ft. and 4 ft., so that if this be one of the animals that were cast adrift in the river as stated, their growth must be very rapid. For my part, I hope it will be proved that it is, and I am sure few of us have any wish to discover that the Yangtsze is inhabited with any reptiles so formidable as alligators or crocodiles.

W. B. R.

Date Unknown

We must "go from home for news." Whence or how, we wonder, did the *Daily Telegraph* get the following. Certainly not from the *Peking Gazette*, so far as we know anything of its contents. Perhaps this cock and dragon story has arisen, however, from the discovery of an alligator in the Yangtsze near Chinkiang last summer:—

"From the *Peking Gazette* we have news of a really wonderful creature in the possession of a fortunate Chinaman at Shanghai. It is described as a veritable river dragon, eighteen feet long, and it has, we are told, the head of a fox, two claws in front like those of a lion, is of the colour of a toad, and when irritated roars with a dreadful sound. In point of appetite it is positively voracious, for when a chicken was handed it for breakfast it swallowed the bird at a gulp. Very naturally the Chinese by whom it was captured show great veneration for the monstrosity, and flock in numbers to worship it. The only question which remains is what shall be done with this remarkable denizen of the water. To transport it to England would be a work of supererogation; there is no necessity to introduce anything into our rivers at home that will destroy the fishes, the present practice of allowing the waste waters from factories and the sewage of towns to flow into our streams being quite sufficient to prevent anything bordering on an excess of vitality. Even the Brighton Aquarium could scarcely accommodate a monster eighteen feet long which takes down fowls at a gulp, and to let the creature loose again upon the ocean would be a sad and dangerous experiment. Yet it is clear that such a wonder ought not to be lost sight of. Who knows the mysteries that might be unravelled by a phenomenon which closely allies fishes, lions, and toads! Clearly the advocates of the evolution theory ought to lose no time in possessing themselves of the animal."

North China Daily Herald

Date Unknown

ALLIGATORS IN CHINA:

THEIR HISTORY, DESCRIPTION AND IDENTIFICATION.

READ BEFORE THE NORTH-CHINA BRANCH OF THE ROYAL ASIATIC SOCIETY, ON 18TH DECEMBER, 1878.

The largest tortoises, alligators, crocodiles, fishes and turtles are produced in the waters.

The alligator-skin drums are resounding.

BY

A. A. FAUVEL,

IMPERIAL CHINESE MARITIME CUSTOMS,

Bachelier ès Sciences de l'Université de Paris and Honorary Curator of the Shanghaï Museum.

Fig 1. A.A. Fauvel wrote the first serious
Western study of the Chinese Alligator

A Modern Plague of Locusts in Israel

Zvi Ron

A Modern Plague of Locusts in Israel
By Zvi Ron

Beginning in March, 2013, waves of locusts began swarming into Israel, crossing the border from Egypt through the Sinai Desert. The swarm first appeared near Cairo on March 2, and then descended on Israel on March 6. The Israeli Agriculture Ministry sprayed pesticides heavily in the Negev region, the southern part of the country, both on land and in the air. [1] The tens of millions of locusts that swept into Egypt over that weekend prompted the Agriculture Ministry to issue a "locust alert". [2] No preventative pesticide can be used in advance of a locust's arrival, and effective spraying can only occur once the insects are settled on the ground for the night or in the early morning before they fly off, the ministry explained, so there was not much to do until the swarms entered the country. [3] After the initial waves of locusts were for the most part eradicated, a new swarm entered Israel on March 27, during the Passover holiday. [4] The major swarms were decimated by the Ministry of Agriculture and only small amounts were left, which were considered relatively harmless. Locusts were reported in central Israel, for example Tel Aviv, and even as far north as Haifa. [5] By the end of Passover, the locust danger had mostly passed, and no major agricultural damage was reported from the Negev region. Still, the Ministry of Defense is helping the farmers in that area destroy the remaining locusts, in order to prevent losses. [6]

The last time Israel fought off a locust plague was in 2005, and swarms of locusts are not that unusual in the region. What made this swarm particularly intriguing was its timing, right about the time of Passover, celebrated this year from the evening of March 25-April 1. This led to many comparisons in news outlets comparing the current swarms to the eighth plague of locusts that struck Egypt before the exodus. [7] Though the timing is uncanny, researchers note that the current plague is a normal ecological phenomenon, and it was not interpreted locally as a form of divine punishment. In the Middle East, locusts typically swarm every 10 to 15 years, and although the pattern can be unpredictable, certain natural factors were identified as contributing to the swarming. In this case, a very rainy winter caused excessive vegetation growth in the region, which resulted in a boom in the locust population. [8] Although some Christian websites pushed the idea that this is somehow an apocalyptic sign [9], no parallels to this appeared in Jewish rabbinic forums, which if they mentioned it at all urged inquirers not to panic. [10] In Israel, although farmers were concerned, the general public was amused by the odd coincidence in terms of timing and the opportunity to taste locust.

The Bible (Lev. 11:20-23) permits certain types of locusts to be eaten. Former Sephardic Chief Rabbi Ovadia Yosef ruled that these locusts, part of the *Acrididae* family that includes grasshoppers, are in fact kosher and may be eaten, but only for those who have a tradition of eating them. Over many centuries, most Jews stopped eating locusts for cultural and sociological reasons, and the tradition of exactly which kind may be consumed was forgotten, and there is virtually no record of the kosher status of locusts in religious texts. However some, such as Yemenite Jews, who used to eat them skewered on shish-kebabs and baked with a light sprinkling of salt, maintained the tradition of which ones are permitted. Some adventurous chefs took advantage of the swarms to prepare various locust centered delicacies. [11]

It should be noted that although the plague of locusts indeed preceded the Biblical Exodus, it was the eighth plague out of ten and occurred some time prior to the Israelites leaving Egypt (Ex. 10:1-20), thus there is no inherent connection between the plague of locusts and the date of Passover.

REFERENCES

1. "3rd wave of locusts from Sinai takes Israel by swarm", Jerusalem Post, March 11, 2013.
2. "Israel mobilizes as millions of locust descend", Jerusalem Post, March 6, 2013.
3. "Agriculture Ministry: More locusts swarm into Israel", Jerusalem Post, March 13, 2013.
4. "Israel battling new swarm of locusts", Jewish Telegraphic Agency, March 27, 2013.

5. "Locusts arrive in Tel Aviv, northern Israel", ynet news, March 9, 2013.

6. "Ministry of Defense Joins Battle Against Locusts", Israel Hayom, May 9, 2013.

7. "Bible comes to life as locusts swarm Israel - Israeli Jews celebrating Passover will easily relate to their ancestors this year – the country has been swarmed by millions of locusts, one of the 10 plagues visited on the Egyptians", Christian Science Monitor, March 27, 2013. "Fertile locusts swarm into Israel on Passover eve arriving with Biblically resonant timing, latest arrivals are ready to reproduce", Times of Israel, March 24, 2013

8. "The Bad News: Just in Time for Passover, Plague of Locusts Arrives in Israel", American Friends of Tel Aviv University, March 25, 2013. The current swarm was reported as "a powerful testimony to the world that the Passover story is real and not legend", "Locusts!", Israel Today, April 2013.

9. "A Million Locust swarm hits Israel, Passover plague Fuels apocalyptic fears are the ten Biblical plagues being imposed on Egyptians by God", Count Down to Zero Time, March 8, 2013. The following message from the founding editor of Breaking Christian News, Steve Shultz, was posted on their website on March 4, 2013, "Locusts in Egypt and even into Israel brings to mind the plague of the locusts in Egypt prior to Exodus and also prior to the first Passover. Now, this plague in Egypt is also just prior to Israel's Passover. Jesus said there would be "signs in the Heavens and Earth" right up until the end of the age. The reader may want to follow this story to see if this sign reveals any insight. Note that even the secular news media is making a connection with the Biblical accounts of the plague of locust and the first Passover."

10.
http://rabanim.net/?nav=send&searchnow=%20%EE%EB%FA%20%E0%F8%E1%E4%20%E1%E9%F9%F8%E0%EC

11. 3rd wave of locusts from Sinai takes Israel by swarm", Jerusalem Post, March 11, 2013.

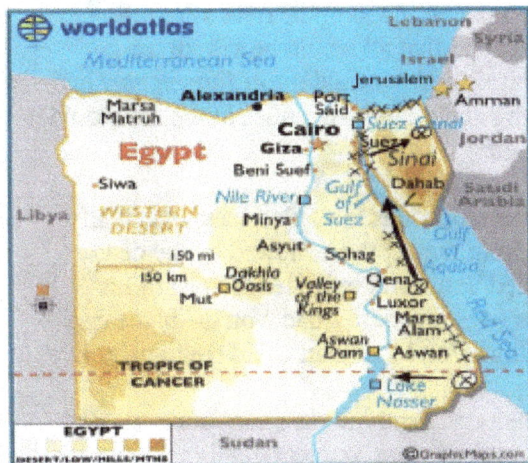

Fig 1 Map of Egypt and E. Mediterranean region showing "invasion" route of locusts. Provided by Zvi Ron

Bornean Crocodile Folklore

Carl Marshall

The *Orang Sungei* (river people) know many legends featuring the crocodile. In one of these tales, related by the elders of Bukit Garam, a peaceful co-existence is claimed between the *Orang Sungei* and the estuarine crocodile. This myth also credits these giant reptiles with the ability to speak.

While staying at the Kinbatangan river I was told the following story: *Terrunggari* - the giant white crocodile.
The *Orang Sungei* speak of a giant white crocodile named *Terrunggari* that once lived in the Kinabatangan river. One day *Terrunggari* grabbed a local boy while he was swimming and dragged him to the mouth of the Kinabatangan, where he told the boy he wanted him to witness a duel he was about to have with another massive crocodile named Berlintang. Berlintang ruled the sea, but he also wanted control of the rivers and *Terrunggari* was all that stood in his way. *Terrunggari,* the white crocodile told the boy that if he saw red blood in the river after the battle, it would mean he had won and that all would be safe in the Kinabatangan, but if he saw white blood it would mean *Terrunggari* had been defeated and that *Berlintang* and all the crocodiles of the sea would rise up, attack and devour the *Orang Sungei.*

The battle between these massive crocodiles took place, and eventually the boy saw red blood float up from the depths, followed by the dead *Berlintang.* The boy then returned home and informed the rest of his village that *Terrunggari* and his river crocodiles had won the battle and the right to stay in the Kinabatangan, and that man could continue to live in harmony with them.
Of course this charming story belongs firmly in the realm of mythology as early travellers reported that attacks were frequent in this area and are occasionally still reported to this day.

Fig 1 Crocodile at Kinabatangan River

Fig 2 Crocodile at Kinabatangan River

Carl's Fortean Spider Experience in Borneo

Carl Marshall

After a tediously long flight from Heathrow via Dubai we finally arrived in Brunei; from there we travelled by four wheel drive into Tanjung Simpang Mengayau, in the state of Sabah, also known as the tip of Borneo. We stayed here for several days to acclimatise to the unfamiliar conditions before we travelled back further south into deeper, more impenetrable jungle. Even though we didn't see much large wildlife in this area of Borneo we did hear that sunbears are sometimes found here but we were not fortunate in seeing any! We did find some very impressive invertebrates one of which, on our return, proved to be a rather interesting find and our experiences with this species were truly fortean in nature.

On our second night at camp, just after a long thunderstorm, Andrew Jackson, my colleague from the Butterfly Farm, was quietly playing the David Bowie classic *A Space Oddity* on his I pod (note this for further reference in the following paragraph), while I was reading Bernard Heuvelman's authoritative work, *On The Track Of Unknown Animals*, when my concentration was broken by a familiar movement to the left of my peripheral vision. It was a fairly large orange sparassid spider of the pantropical genus *Heteropoda* running at high speed across the floor in the direction of our makeshift kitchen, and only when it finally paused could we briefly view it more clearly. It was a relatively large specimen about 2.5 - 3.0 cm across, with a pale cephalothorax (fused body and head) and abdomen covered with distinctive orange hairs that were also on each of the spiders eight appendages. It had dark chelicerae (mouthparts) and a dark dorsal stripe running vertically from the spiders eyes all the way down, almost to the tip of the abdomen. We viewed it for about three minutes before it disappeared out of view into a crack, and after leaving it alone we thought nothing more of it until we arrived back in the UK when we started identifying encountered species.

It seems that our Huntsman spider most closely resembles a newly discovered species known taxonomically as *Heteropoda davidbowie* -

- yes a species of spider has been formally named the David Bowie's spider! - apparently named after the rock star in an attempt to raise public awareness of the increasing number of arachnid species facing extinction. Our encounter with this spider might also be of importance as *H. davidbowie* is known from western (peninsular) Malaysia, Singapore, Indonesia, Sumatra and maybe also southern Thailand and to my knowledge is *not* known from anywhere on the island of Borneo.

Within my profession I have worked with a variety of species from the genus *Heteropoda*, studying many live specimens and after comparing what we witnessed at camp with the formally identified Bornean species, I am now contemplating the possibility that *H. davidbowie*, if its numbers do not drastically decline, might also be discovered in Borneo.

With our appetites whetted for further exploits, next year Andrew and I return to Borneo on the hunt for giant pythons and also Borneo's answer to Orang pendek, the equally elusive Batutut but while there we shall certainly be searching for this mystery arachnid.

Borneo's Mystery Animals - Report of a Visit in March 2013

Carl Marshall

Borneo is the third largest island in the world and the largest island in Asia. It is located north of Java, west of Sulawesi, and east of Sumatra. At 130 million years old, Borneo has some of the oldest rainforests in the world. There are about 15,000 species of flowering plants with 3,000 species of trees of which 267 species are *dipterocarps*. Borneo has 221 species of terrestrial mammal and 420 species of resident birds. There are about 440 freshwater fish species which is about the same as Sumatra and Java combined, 149 of them are endemic to Borneo.

The Borneo rainforest is one of the only remaining natural habitats for the Borneo orangutan and is an important refuge for many endemic forest species, including the Asian elephant, the Sumatran rhinoceros, the Borneo clouded leopard, the Hose's civet and the dayak fruit bat.

In March 2013 my colleague and I from the Stratford upon Avon Butterfly Farm travelled into the interior of Malaysian Borneo to study its ecology and biodiversity and were lucky enough to encounter many endangered species such as the forest elephant, the slow loris, pig-tailed macaaque and many unique chiroptera species.

Not only were we looking for any evidence of rare, known fauna, we were also inquiring after cryptozoological species. The list below contains some hitherto undocumented reports.

[1]Giant saltwater crocodile: A soldier informed us of a colossal 35ft crocodile he had witnessed near Lok Batik, Sabah.

[2]Giant reticulated pythons: Our guide from Ulu Kamanis informed us of giant pythons 30ft +. More research on these in 2014.

[3]luminous paradise type birds: We were informed by Matthew Lazenby of glowing paradise type birds in the deep forests of Ulu Kamanis.

[4]Giant black orangutan: We were informed of Indonesian giant black orangutans - more on these in 2014

[5]Sabah sky rods: Matthew Lazenby has done much research on the flying rod phenomenon. He has taken part in a reconnaissance of a deep cave, that after viewing back the video footage, shows very strange flying objects that Matthew claims have different flight patterns to moths filmed using extremely short exposure times. We will meet with Matthew (Jigger) again in 2014 and shall hopefully have further updates on this strange phenomena.

[6]Batatut: The Bornean orang pendek is named the Batatut. I found no evidence of this cryptid from Malaysian Borneo! Further research on this creature in 2014 expedition to Indonesian Borneo.

[7]Possible OOP *H.davidbowie*: While staying at Tampat Du Aman we briefly witnessed a *Heteropoda sp* that closely resembled *H. davidbowie*, a species supposedly only found in peninsular Malaysia - could this species also be found in Borneo? We will investigate this further in 2014.

[8]OOP long arm scarab beetles: Max Blake of the CFZ informed me last year that because of the distribution of many species of Euchirinae, specimens might be found in the middle east and incredibly on Borneo. No evidence as yet! More research on this in 2014.

This list will no doubt be updated in 2014 when we return to Borneo for our follow up expedition - next time mainly researching Indonesian Borneo.

Fig 1 Pygmy elephant © Carl Marshall

Fig 2 Southern pig tailed macaque *Macaca nemestrina* © Carl Marshall IMG 1893

Fig 3 Female probiscous monkey with baby *Nasalis larvatus* © Carl Marshall IMG 2027

WERE SOME FLYING SNAKES REALLY PEACOCKS?

Richard Muirhead

Fig 1 Flying snake as depicted by Crispjin de Passe, early 17th Century.

Fig 2 Head of a peacock .

Wikipedia Creative Commons.

I wish to talk here about the possibility that some flying snakes were actually peacocks, rather than pheasants as suggested by Dale Drinnon,although to be fair on Dale he did mention peacocks.(1) I am not ruling out the possibility that *some* flying snakes were really pheasants mistakenly identified as flying snakes,but I am suggesting that the possibility that the flying snakes reported to inhabit the woods and culverts around

33

Penllyne Castle in Glamorgan, Wales were very probably pet peacocks which had at one point in time escaped from the immediate grounds of that Castle. After all the ring necked pheasant, of the kind normally seen in Wales doesn't have the upstanding feathers on its head as in the flying snake depicted in `America` by Crispijn de Passe (see page 37) as does the peacock, nor would there be anything exceptional and surprising about a rustic person in late 19th /early 20th South Wales seeing a pheasant ,so why make any comment on it to visiting folklorists such as Marie Trevelyan who made the be-jeweled "snakes" popular amongst contemporary cryptozoologists? The pheasant has been known in Wales since Tudor times, 1485-1603.(2) "In Glamorgan the Pheasant was evidently scarce or unknown until towards the close of the eighteenth century, since in 1781 Thomas Mansel Talbot was awarded a gold medal by the Glamorganshire Agricultural Society for his spirited endeavours to introduce the English Pheasant into this county…"(3) It is a well known fact that even up to the present day peacocks have been kept as attractions in the gardens of British country estates and it is by no means beyond the bounds of possibility that during a storm,or civil unrest,or deliberately for one reason or another a number of these wonderful peafowl escaped and even bred. Pheasants may have been kept and raised as game. Admittedly, officially there never have been peafowl in N,America. But could they not have been transported there as pets or food? The first Welsh are rumoured to have reached North America in the form of the Welsh prince Madoc in 1170. He founded a colony which mixed with Native Americans. Interestingly, the painting or engraving `America` mentioned above,shows the flying snake/"peacock" in its proper context, that of just one animal amongst many. Does this mean that rather than peafowl ,flying snakes were common in North America in the 1600s?

Between 1541 and 1556 the Italian naturalist Hieronymus Benzo wrote in his journal during a visit to Florida :" I saw a certain kind of Serpent which was furnished with wings, and which was killed near a wood by some of our men. Its wings were so shaped that by moving them it could raise itself from the ground and fly along, but only at a very short distance from the earth." (4)

Peacocks can only fly short distances. Also, remember the long snake-like tails of peacocks.

On May 7th 2013 I posted the image on page 37 on the `Cryptozoology` Facebook Forum with the question: " Has anyone worked out what the "flying snake"/basilisk here could be (in Crispijn de Passe`s America)? It is usually not seen in its context with other animals commonly known." Here are the most relevant responses:

Stuart Paterson: It`s a 16th century mannerist[1] painting by van de Passe, idealising the American continent. Therefore, the bestiary is partially imagined, partly based on the reports of others & partly indicative of the sort of things imagined to be there.

Adam Kukoleck: How about it being a once commonly known basilisk?

Victor Vasquez: Because they look like they may be cannibals of some sort I would imagine that they are a European representation of Central American Natives such as the Aztec. The flying serpent under the foot might represent Quetzalcoatl or even Lucifer since they were seen as evil for their blood sacrifices. Maybe even a reference to Isaiah 30:6 .That Bible quote was often used against a people to say their riches do not belong to them & can be taken away because they are evil & against God.

I find the imagery some sort of propaganda against the First People of the Americas.

Martin Kilmer-True: To be fair, the Aztec empire WAS absolutely batshit insane.

Mike Richardson To be boring , what if it was just the remains of a bat? I could see how a description of a bat plus some remains could be inspiration for a winged snake.

---

[1] Mannerist : An artistic style of the late 16th century characterized by distortion of elements such as scale and perspective.

Richard Muirhead: Have there ever at all been any reports of peafowl in America?

Stuart Paterson: Not till the late 19thC. However, as I said earlier, European mannerist painting at that time imagined animals and `cryptids` in little explored lands. I suggest this is that. What's more intriguing is Gauguin's mystery bird, some sort of rail perhaps, as he painted at Hiva Oa.

Rowson Zhen: I believe that is exactly what it is depicting, a snake with wings.Ancient historians quite often depict dragons as reptiles, especially as pythons and boas but a few have mentioned "dragons with wings" .If you were to look at the "evolutionary record" of snakes, you'd find that prehistoric snakes came in a plethora of forms and possessed different physical traits. Some had legs, some had flippers (specimens of both have been discovered) and these ancient historians mentions them as well .Along with this, there are varieties of ones with wings. Unfortunately, snakes do not fossilize easily so much of the fossil record remains incomplete and there are many missing links and branches depicting their lineage. But it is quite conceivable that some might have wings, for either gliding or flying. Btw, it is also worth noting that there has been and still exists snakes with beaks.

Rowson Zhen: Even contemporary snakes have horns,tentacle protrusions, leaf like protrusions, etc. I wouldn't doubt it at all of prehistoric snakes having feather like protrusions. As of now, only a few species of contemporary snakes have beaks and all are ocean dwelling so they don't have feathers.

John DeSilva: I believe the quetzal (from Central American folklore) was described as a feathered flying and is in actuality a long tailed bird...

Fig 3

Crispijn de Passe's `America`

REFERENCES

(There are reports of feral peacocks in Britain, see:
http://www.thewatchforum.co.uk/index.php?showtopic=79312 about a
case in December 2012 in Cambridgeshire. On December 4th 2007 Wales
Online reported that peafowl were escaping into Cardiff from Cardiff
Castle.)

1. Dale Drinnon. Flying Snakes Parts 1 and 2 in Flying Snake vol 1no 1
pp 20-23 and vol 1 no 2 pp 13-17.
2. The Pheasant in Wales
http://www.britishbirds.co.uk/search?model=pdf&id=3742
3. Ibid.
4. Yahoo question and answer
forumhttp://answers.yahoo.com/question/index?qid=20110823054106AA
H9bLj

SOME NOTES ON THE DEER OF HONG KONG

RICHARD MUIRHEAD

Fig 1 Spotted deer (Axis deer) in a zoo on Lantau Island from The Star newspaper of July 29th 1977. I am sorry about the poor quality of the above

In the 1860s, axis deer were introduced to the island of Molokai, Hawaii, as a gift from Hong Kong to King Kamehameha V. Today the deer are found plentiful on Lanai, another of the Hawaiian Islands. The Paniolos (cowboys) were instructed to lasso the deer from Molokai and to bring them for shipping to Lanai. Hawaii wildlife officials believe people have flown the deer via helicopter and transported them by boat onto the island. (1) I have never seen this mentioned in any standard book on Hong Kong's fauna, past or present.

According to Introduced Mammals of the World `Axis deer `(Chital) from India have been introduced and established in the Hawaiian islands (Gottschalk 1967; Tomich 1969.) In 1867, seven axis (three males and four males) were shipped to the reigning monarch of Hawaii as a gift from his envoy in Japan. Some of these came from the Upper Ganges River.India, but some died on the voyage and were replaced at Hong Kong *with animals of unknown origin* (2) [Emphasis my own.]

Jon told me on June 8th that within the last 10 years some have come to believe that there are two species of muntjac in Hong Kong, Reeves muntjac and the Indian muntjac. This is interesting given the possible importation during the days of the East India Company (1600-1874) of the axis deer (see above) or Chital deer from India,Bangladesh,Sri Lanka,Nepal or Bhutan to Hong Kong sometime between the early 1840s and 1860s. This is my theory anyway. The image on the previous page is of a chital, found in the Hong Kong newspaper The Star on July 29th 1977. There was a small herd of the deer on Lantau island then. So was there a continual population overlooked in Hong Kong for about 100 years?

The Proceedings of the Zoological Society of London for 1864 mentions the following:

"In the gardens of Messrs. Jardine,Matheson,and Co. In Hong Kong I saw several bucks and does of C.sika and C.taivanus, and also of C.axis in winter dress. The bucks of the two former had manes about the neck;C.sika was spotless, C.taivanus with indistinct spots, while C.axis was of a rich yellowish-brown colour, with distinct white spots. The latter had long,thin,reddish tails,and,I think,are identical with the true C.axis. They are from Hankow,interior China. The bucks of C.sika, otherwise similar, differed a good deal in size; they were, I believe, from different islands of Japan, the smaller from Nippon, the larger from Yesso.

"The Deer from China may therefore be enumerated, as follows:-

Cervus dama, L. In gardens at Canton.

C.axis, L. From Hankow, Central China.

C elaphus, L. Summer Palace Gardens.

C wallichii. Tartary, beyond the Great Wall (horns seen by me at Peking)

C swinhoii, Sclater. Island of Formosa

C hortulorum , Swinhoe (" C pseudaxis"? , Gray). Gardens of Summer Palace.

C taivanus, Blyth. Island of Formosa.

C sika. Island of Japan.

C mantchricus. Mantchuria (sic) Size larger than C taivanus, with horns short in stem, and more resembling those of C sika Colouring very similar (in winter coat) to that of C taivanus. Red patch on occiput, on each shoulder, and on side of neck. Black line down back somewhat indistinct; mane from side and back of neck rather long,thick, shaggy,and dark-coloured. Belly pale reddish white. Thighs light reddish brown." (3)

REFERENCES

1. Wikipedia http://en.wikipedia.org/wiki/Chital

2. J.Long Introduced Mammals of the World (2003)

3. Proceedings of the Zoological Society of London 1867 p. 169

Horned Rats

Photo.]   [*Thomas Hunt.*

Horned Rat Killed near Evesham.

In the Spring of 2013 Bob Skinner provided me and the Fortean
zoological community in general some interesting data on a
cryptid new to me - the Horned Rat. The following pages consist
of all I know about this creature between 1857 and 1912 after
which the trail seems to grow cold. The 1857 account is from the
U.S (via a French paper)the **Janesville Morning Gazette**
March 11th 1857 p. 12:

"Zoological- Did you ever hear of a horned rat? Last Monday
the police court was occupied by a case brought by a scientific
man against a Zouave[1] for selling him two horned rats, and
cheating him out of twenty dollars. As the prosecutor gave in his
evidence with singular talents, you must let me give the column
to him" Messieurs," said the savan, addressing the court, being
first duly sworn on the Holy Evangelists to tell the truth..."that
man(pointing to the Zouave at the bar with his finger) has
cheated me most rascally. That fellow knows I am a student of

[1] French light infantry in N.Africa 1831-1962

botany, mineralogy, geology, conchology,anthropolity, and zoology;and one day he came to see me. Says he `Monsieur, did you ever see a horned rat?` `A horned what?` says I. ` A horned rat` says he. `No` says I . ` and what`s more, I`d like to see the man who had, for neither Buffon, nor Cuvier, nor Lascepede say anything about such an animal. There is not one at the garden of plants, and - he interrupted me, Messieurs, to say `I have got one.` ` Go to the devil,` says I `Come and see for yourself,` says he; `It`s alive and *piert* as a young rhinoceros.`

On we both went , and sure enough in his house I saw a big fat rat, which had on its nose a fur-covered excrescence, which was vertebrated,and - most wonderful of all - the horn was larger at the at the top than at the bottom. The prisoner at the bar took the rat out of the cage, held its mouth, and placed it in my hands. I plunged a pin into the horn, the rat jumped and screamed, blood issued from the wound. `Bon`, says I, `it`s no humbug; the horn is really a part of the body.` I paid fifty francs for the rat and carried it home...I happened to meet an officer who had served long  in Africa, and I said to him: `Mon General, you have served a good many years in Africa`. `Alas for me`, says he. `Did you ever hear of horned rats` , says I. `La`, says he `haven`t I; they are as plenty at Algiers as grisotts in Paris`. - `Then`, says I, ` be good enough to explain to me why their offspring never have horns` I told him my story. When I got through I thought he would have died a laughing; he shook , he rolled on the sofa, he cried. `Bon` says I, `I`ve been cheated.` When he became calmer I asked him to explain. `Horned rats,` says he, `are manufactured by the Zouaves, who take two rats, puncture their noses, and graft in each rat`s nose, a portion of the other rat`s tail ,and when the tail adheres to the flesh of the nose the tails are cut, and each rat has a horn" **Paris Cor N.O.Picayune."** (1)

Northern Argus (Clare, South Australia) 9 May 1873 p.3

A MAKER OF MONSTROSITIES

Paris has just lost another celebrity, one Femorus, whose skill in the manufacture of monstrosities must have made Barnum envious. Femorus first tried his apprentice hand on beasts, and succeeded in concocting no end of two-headed cows, five legged birds, horned rats, and such harmless wonders, which went the round of all the fairs for many years. These, however, at length got stale and unprofitable.Modern civilisation demanded something more attractive, so Femorus turned his cunning to the human race itself. In 1854 he was sentenced to five years imprisonment for trying to implant the wings of a swan in the back of a child two years old with the object of making a second cupid! On leaving the prison he recommenced business by manufacturing all sorts of "natural curiosities" , too ugly to specify, anatomical museums, but the demand not being equal to the supply, he was forced to brush up his ingenuity once more. This time he resolved to operate on himself, and he attempted to engraft the comb of a cock - a Gallic one, of course - on his own head. It was doomed to be his last wonder. The engrafting resulted in abcess, from which he died after a fortnight's suffering. At least, so the Paris papers tell us. - **Echo** (2)

Wellington...Mr Hookway of Marian's Buildings, has added a very peculiar curiosity to his household possessions a horned rat. I hope it is the only one of its species. Rats I detest, but horned ones would be worse. I have not seen the animal. Which is being stuffed by Mr Palfrey, but I understand its top front teeth have grown to the back of its mouth, and one of them coming through,has grown and formed what appears to be a horn, very like a ram's horn. Well I'm very glad indeed Mr. Hookway caught it and hope we shall not be favoured with any more of its kind,or Wellington will not be an ideal dwelling for one of its community. **Western Times** April 25th 1902 p. 12

The following is the text accompanying the photo on page 41:

The Countryside Magazine, Vol 3 p. 326, 28th April ,1906

A Horned Rat. - One's natural impulse on seeing such a photograph as this of a stuffed rat, is to suppose that the taxidermist has played a practical joke; but the following correspondence regarding it justifies our publication of the picture. The first notice of the creature came to us on January 10th from Mr, B,Taylor,Arrow, Alcester, as follows:-

"At Harrington, near Evesham, during early part of December last a doe rat with two solid horns slightly curved measuring 1 ¼ inch long and 5/8 in round at base, was killed. Must have been a formidable creature."

Next on January 24th came the following from Mr Thomas Hunt,naturalist, Church St, Alcester: - " Have forwarded you a photo of a most curious horned rat killed at Norton, near Evesham. I should think it would be very interesting in your COUNTRY-SIDE, which I often take, and should be glad to hear from you."

Further enquiry brought this on February 10th from Mr T.Perry,Norton, near Evesham, who killed the rat: -

"Dear Sir, - Of course the rat had the horns on when alive as they were on when I killed it at Harvington,near Evesham , on November 10th, 1905. And we have seen a photo of it and it`s very good indeed . Just like a rat now it`s preserved, - Believe me, yours truly, T.Perry.

P.S - there were more men working with me at the time it was killed and several saw it before we had it preserved.

Washington Post March 26th 1907 p. 6

A Kansas man asserts that he recently saw a rat with horns. As the authorities insist that the prohibition law is being enforced in Kansas, there must be something wrong with the soda water in that State.

Same or similar paragraph also in

Piqua Leader-Dispatch (Ohio) April 1st 1907

Ukiah Dispatch Democrat (California) April 3rd 1907

Attica Fountain Warren Democrat (Indiana) April 4th 1907

Ardmre Daily Ardmorite (Oklahoma) April 8th 1907

Clinton Plain Dealer (Indiana) May 3rd 1907 p. 7

Checotah Times (Oklahoma) May 10th 1907

Milford Mail (Indiana) November 3rd 1907

Des Moines Daily News (Indiana) May 12th 1912 p.12:

AND KANSAS IS DRY

A Kansas man asserts that he recently saw a rat with horns.

UNIDENTIFIED SUBMARINE OBJECTS: CASE-STUDIES FROM AROUND THE WORLD

NICK REDFERN

Unidentified Submarine Objects: Case-studies from around the world

By Nick Redfern

When Richard Muirhead asked me if I would be interested in submitting a paper – on the subject of Unidentified Submarine Objects (USOs) – for the latest issue of his *Flying Snake* magazine, I didn't waste any time in saying 'yes'. And for one very good reason: I have many such reports in my files demonstrating that the oceans of our world are deeply weird places, indeed. It's probably fair to say I have enough cases on-file to write a complete book on the subject, but space issues mean I'm going to limit myself to just a few of the most intriguing USO-based accounts and incidents I have on file.

Bloop: beast, unknown craft, or a fuss about nothing?

During the 1960s, the U.S. Navy created a vast array of underwater microphones, or hydrophones, around pretty much the entirety of the planet. The reason was as simple as it was deeply secret: to keep careful track of the movements of the Soviet Union's fleet of submarines, many of which were armed with sizable numbers of nuclear warheads. The project became known as SOSUS, or Sound Surveillance System.

To this very day – even though the Cold War is now long over and done with – the listening stations are still in place, hundreds of meters below the surface of the world's oceans, at depths where sound waves become trapped

45

in a layer of water known as the Deep Sound Channel. It's here that temperature and pressure cause sound waves to keep moving without being wildly scattered by the ocean surface or bottom. As for those sounds detected by SOSUS, most can be traced back to everyday things like whales, ships and even earthquakes. It's much the same for the National Oceanic and Atmospheric Administration (NOAA), an agency within the U.S. Government's Department of Commerce. Back in the summer of 1997, however, things were far from being everyday in nature. NOAA's Equatorial Pacific Ocean autonomous hydrophone – which was created to augment the Navy's SOSUS program and plays a key role in monitoring populations and migrations of deep-sea animals – detected a strange anomaly at a remote point in the South Pacific Ocean, west of the southern tip of South America. It was an anomaly that suggested deep in the oceans huge, terrifying monsters may lurk about which the U.S. Government knows a great deal. Or didn't suggest that, depending on whose version of events you accept as being valid.

Within official circles, the anomaly became infamously known as Bloop, and according to NOAA's records was detected quickly in frequency for around sixty seconds and was of enough amplitude to be detected on sensors ranging over 5,000 km. Not only that, the sonic frequencies that were monitored convinced some within officialdom that if a living, breathing thing really was roaming the waters off South America, then, as the recordings suggested, it was clearly bigger than any recognized or categorized creature of the sea. Was a marauding monster of massive proportions on the loose? Had H.P. Lovecraft's terrifying, giant-octopus-like Cthulhu coiled its way out of the realm of horror fiction and into the world of reality?

One of the first suggestions put forward by those that felt, or suspected, the Bloop recordings just might be evidence of a huge beast swimming deep in the oceans of our world was that it was a giant squid. And, by that, we're talking the giant of all giants. Certainly, squid can grow to impressive sizes: current estimates are between thirty-three to forty-six feet, although unverified reports of sixty foot long monsters do exist. Might Bloop have been an even bigger beast, the definitive granddaddy of the squid world?

Certainly, within the domain of adventurous fiction, giant squids grow to massive sizes, such as those that took on Captain Nemo and his crew in Jules Verne's classic novel *Twenty Thousand Leagues under the Sea*; the one that caused death and mayhem in *Beast*, a novel by *Jaws* author Peter Benchley; and that which provoked terror for the white sperm whale-obsessed Captain Ahab in the pages of Herman Melville's acclaimed *Moby-Dick*. Fantasy is one thing; but what about reality? Comments and observations – some controversial and others less so – were soon forthcoming.   Phil Lobel, a marine-biologist at Boston University, Massacusetts, was skeptical of the giant squid theory. He noted that cephalopods lack gas-filled sacs, thus preventing them from making noise. But Lobel didn't dismiss the fact that something occurred. Rather significantly, he did admit that, while in his opinion the source of the strange recording was not a huge squid, it probably was something biological in origin.

Christopher Fox, of NOAA's Portland, Oregon facility, confirmed that other such anomalies had been recorded over the years, and across the planet, and had been given a variety of names, such as Gregorian Chant, Upsweep, Slowdown, Whistle, and Train. Interestingly, Fox did not discount the possibility of the Bloop signal coming from a living creature, as its signature was, he explained, somewhat akin to a fast variation in frequency to the sounds made by known and identified animals of the oceans.

Nevertheless, from the perspective of NOAA, at least, the matter has now, finally, been laid to rest. The official stance, today, is that the mysterious sounds recorded in 1997 are not unlike what one would expect to encounter in the fracturing and cracking of large icebergs. NOAA adds that the iceberg they believe caused all the fuss back then was most likely situated somewhere between Brasnfield Straits and the Ross Sea, or possibly at Cape Adare, a well known source of what are termed cryogenic signals.

And that's where matters presently, and officially, stand regarding the Bloop signal of 1997. Unless, of course, you subscribe to the theory that, perhaps, H.P. Lovecraft's fictional Cthulhu is not quite so fictional, after all. Or that the U.S. Government is hiding the truth about terrible, marauding beasts that lurk in the deep, ancient waters of our planet.

Terror at the cove

In 1997, a strange event occurred in the English west-country that had at its heart a definitive USO angle. It all began on October 1, 1997, as Nigel Wright of the Exeter Strange Phenomena Research Group reveals:

'Approximately three weeks ago two young men were swimming in Otter Cove [, Lyme Bay, Exmouth]. As darkness drew in, they decided to make for the shore and change to go home. As they got changed, one of them looked out to sea. He saw what he described as a "greenish" light under the surface. He called to the other young man and they both watched as this light "rose" to the surface of the water. The next thing they knew there was a very bright light shining into their faces. They turned the scene and fled.'

Meanwhile, on the top of the cliffs, equally strange things were afoot. The two young men raced for the car of a relative and breathlessly explained what had happened. Incredibly, she, too, had seen something highly unusual in in precisely the same time frame on the road leading to Otter Cove: a strange animal that she likened to 'an enormous cat'. Whatever the origin of the beast, however, she was certain of one thing: it was, to quote her, 'all lit up' – glowing almost.

On the following day, a dead whale was found washed upon the beach below the cliffs. This did not appear to have been merely a tragic accident, however. On receiving reports that a whale had been found in precisely the area that anomalous lights and a strange creature were seen, Nigel Wright launched an investigation.

'The first thing that struck me as I looked on at this scene,' recalls Wright, 'was how perfect the carcass was. There was no decay or huge chunks torn from it. Then, as I wandered around it, I noticed that there was only one external wound: in the area of the genitals a round incision, the size of a large dinner plate, was cut right into the internal organs of the mammal. The sides of this incision were perfectly formed, as if some giant apple corer had been inserted and twisted around; from the wound hung some of the internal organs. I quizzed the official from English Heritage, who was responsible for the disposal of the carcass. He informed me that no natural predator or boat strike would have caused this wound. As I looked at this sight, the first thing that came into my mind was how this looked just like the cattle mutilation cases of recent times.' Wright was also able to determine that this was not the only time that unusual lights had been seen in the vicinity of Lyme Bay.

'No precise date can be given for the evening when a fishing boat encountered a strange light over Lyme Bay,' explains Wright, 'but, since this was told to me by the skipper of the vessel concerned, I can vouch for its authenticity. The vessel in question was five miles off Budleigh Salterton. The crew became aware of a bright, white-blue light which hovered some distance from the boat. At first they thought it was a helicopter but they heard no engine sounds, nor saw any navigation lights.'

Wright was told by the captain of the vessel that the night had been 'bright and clear' and that if the object had made any noise, it would certainly have been 'audible for miles'.

'The light remained stationary for about one and a half hours. Judging by the mast of their vessel, which is twenty-eight feet high, the crew estimated that the light was not much higher than that,' adds Wright. 'It then very suddenly disappeared.' The mutilators, however, did not disappear – far from it, in fact. **USOs and a mysterious island**

Puerto Rico, or to give it its correct title, the Commonwealth of Puerto Rico, is what is known as an unincorporated territory of the United States, located in the Caribbean Sea. And according to some, it may very well be the one place – possibly more than any other on the planet - that is home to not just one secret base, but to an overwhelming plethora of classified locations, certainly of a governmental nature, and maybe even of an alien nature, too. Over the course of the last twenty-years or so, the people of Puerto Rico have been swamped by a multitude of UFO encounters, sightings of strange and unearthly-looking craft surfacing from both mountainous and cavernous lairs, and run-ins with strange, vampire-style creatures that one might accurately describe as the distinctly evil-twin to Steven Spielberg's benign E.T.: the Chupacabras. And then there are the USOs of Puerto Rico.

In 2004, when I visited Puerto Rico for the first time – in search of the Chupacabras - I was told of the account of a former civil-defence employee, who had seen a gigantic, unknown craft rise silently out of the coastal waters of the island, while he was on an early-morning jog in the spring of 1999. In this case, the vast device, which was viewed at a distance of around half-a-mile off the coast, or perhaps slightly more, wobbled slightly – rather like a falling-leaf - as it took to the skies, and then streaked vertically at a fantastic speed, before finally vanishing from view as it grew ever smaller, and was finally lost due to the effects of the bright, rising sun.

Further rumors of a potentially-related nature were also provided to me on that same expedition to the island. They came from a retired police-officer who had heard rumors to the effect that, somewhere off the coast of Puerto Rico – he was not entirely sure where exactly – in late-1993, elements of the U.S. Navy spent several days tracking, via sonar, the movements of a huge USO in the deep waters off Puerto Rico. Perhaps aware of its potentially hazardous nature, the U.S. Navy contingent was ordered to merely carefully log the movements of the undersea craft, but never to engage it any way, shape or form whatsoever that might be interpreted as hostility.

Taking the above into thoughtful consideration, is it truly feasible that Puerto Rico might be home to a massive undersea installation? When one realizes that we, the Human Race, have had the ability to construct such science-fiction-like facilities for decades, then the possibility becomes all-too-real, and not so unbelievable, after all. And, make no mistake: evidence of our very own undersea abilities is far from lacking.

For example, an October 1966 document prepared by one C.F. Austin, of the U.S. Naval Ordnance Test Station at China Lake, California, includes a truly remarkable statement. Titled Manned Undersea Structures – The Rock-Site Concept, it states in part that: 'Large undersea installations with a shirt-sleeve environment have existed under the continental shelves for many decades. The technology now exists, using off-the-shelf petroleum, mining, [51]

Fig 1 Giant Squid, Harper Lee, 1884

submarine, and nuclear equipment, to establish permanent manned installations within the sea floor that do not have any air umbilical or other connection with the land or water surface, yet maintain a normal one-atmosphere environment within.'

If, as this previously-classified U.S. Navy document demonstrates, the government of the United States was constructing undersea installations - with a comfortable shirt-sleeve environment, no less – a number of decades before the documentation was even prepared in the mid-1960s, perhaps someone else, someone from a world far, far away, has secretly been doing likewise. And, maybe, they chose Puerto Rico as their secret base of both underground and undersea operations.

A triangle of undersea puzzles
Extending from Bermuda in the north to southern Florida, and then east to a point through the Bahamas past Puerto Rico and then back again to Bermuda, is a truly ominous realm of wild, churning and turbulent waters known infamously as the Bermuda Triangle, a permanent fixture in the western part of the North Atlantic Ocean, and one that has become renowned for the hundreds of aircraft, ships, boats and unfortunate souls that have disappeared in the area without trace – and for decades, too.

Fig 2
Puerto Rican waters
© Nick Redfern

Down to earth explanations for such vanishings, it goes without saying, most certainly proliferate. Compass malfunctions, disorientation, sudden and violent bouts of severe weather, mechanical and electrical failure, and pilot error are just some of the conventional theories that have been offered as answers relative to why there should have been so many disappearances in such a clearly delineated area over so many years. But, not everyone is quite so sure that those particular theories provide all the clues to solving the maritime mystery; one of the reasons being that on numerous occasions USOs have been seen in the area.

One particularly significant case involving a craft of distinctly unknown origin occurred in April 1973 when a Captain Dan Delmonico, a calm and collected character with a fine reputation for being grounded and logical, had an encounter that could be considered anything but grounded and logical. It was around 4.00 p.m., while negotiating the waters of the Gulf Stream, specifically between Great Isaac Light, north of Bimini, and Miami, when Delmonico was amazed by the sight of a large cigar-shaped object - nearly two hundred feet in length, grey in color, and with rounded ends – which shot through the water, not surprisingly amazing and astounding Delmonico in the process. Who, or what, piloted the strange submersible on that April 1973 afternoon remains unknown.

A real-life X-File

Ninety sixty-six saw an unusual event occur at Pasajes, Northern Spain that caught the attention of the Ministry of Defence. From a radio officer attached to the *S.S. Patrick M. Rotterdam*, came the following, which I found in 1997, in a then-newly-released batch of formerly classified

British Ministry of Defence files on UFOs. This case does not involve a USO directly, but since the witnesses were at sea, it may be of some relevance. A letter sent to the MoD by the ship's captain reads thus:

'Perhaps the following will be of some interest to you or Jodrell Bank. Whilst at Anchor at Pasajes, North Spain on 22 April at 2100 Bst in a very clear sky, one of the crew noticed a bright patch in the sky and drew my attention to it. It appeared stationary and squarish, the area being about 4 times the size of a full moon. Several of the crew watched, being interested and of course at anchor, there is very little to do. The patch elongated and became brighter and to our amazement a complete ring, similar to pictures of flying saucers, bright and distinct with dark centre. For several minutes this object remained visible then returned to a patch, receding elongated again. Then it branched out to form a letter M. When the ring was clear it was about [the] same size as a full moon. We know it was not the moon because the moon was in another quadrant and lying on back at [the] same time. The patch receded away into distance. I can assure you none of us were drunk.' And there you have it: a round-up of just a few of the weirder USO-themed reports from my files, all of which suggest maybe it's not to the stars – but to the seas – that we should be looking for the answers concerning the many UFO-themed mysteries that dominate our world.

Nick Redfern is the author of many books, including *Wildman*, *Monster Diary*, and *The Pyramids and the Pentagon*. He can be contacted at nickredfernfortean.blogspot.com

Sources:
The Rising of the Moon, Jonathan Downes and Nigel Wright, Domra Publications, 1999.

Report by Nigel Wright, September 1998. Exmouth Journal, 16 October 1997.

Letter to the Ministry of Defence, April 26, 1966.

Monster Files, Nick Redfern, New Page Books, 2013.

Keep Out!, Nick Redfern, New Page Books, 2011.

The World's Weirdest Places, Nick Redfern, 2012.

NOTES AND QUERIES

The following appeared in **Knowledge** magazine in October 26th 1883:

IS THE COMMON EARTHWORM LUMINOUS?

Having had my attention drawn last night(October 8) to some phosphorescent streaks on the road,which had been observed by some friends, I sallied forth with two of my pupils,and having found two streaks,struck some matches with which I provided myself, and found that the phenomena proceeded from what were apparently two small earthworms. Is the common earthworm luminous, or is there a special light-giving variety?

They had not the slightest resemblance to glowworms,but were exactly similar, as I have said, to small earthworms. J.J.C. Fenton. (1)

THE FERAL SCARLET MACAWS OF KIRKBY STEPHEN,CUMBRIA

The Guardian March 14th 2013

From the moment the nerve-jangling screeches – on a par with the sound of glass stoppers turning in the necks of bottles – were unleashed from the rooftops, it was obvious which of the shoppers in the street were local residents and which were first time visitors. The former went about their business without an upward glance; the latter stopped in their tracks,stared upwards, then reached for their cameras. Kirkby Stephen's small flock of scarlet macaws has been a raucous feature of the town ever since the late John Strutt began to provide a refuge for unwanted parrots on his farm at nearby Eden Place. As a dedicated animal lover and conservationist, Strutt managed his 900 acres using farming methods for the benefit of birds, butterflies and wild flowers, and also indulged his passion for exotic birds, which he could never bear to see confined to an aviary.

Our first encounter with his feral macaws was on a footpath near his farm, where a few feathers left under the trees by a bird preening left us puzzling over who had been the owner of such exotic plumage, all the primary colours from a child's paintbox. For the most part, the macaws stay close to home in the Eden valley, but they are regular visitors to Kirkby Stephen's rooftops, where they have become local celebrities. There will be some who will point to the proliferation of troublesome, feral ring-necked parakeets in London and frown on the deliberate introduction of any non-native species, but this population has remained small, stable and local.

As we crossed the road to the bakery, a pair peered down at us from the parapet above, technicolour adornments to grey roofs on a grey day. Then they flew low and fast above cars and lorries along Market Street, trailing long tail plumes and ear-piercing screeches – Amazonia on a bitterly cold Cumbrian afternoon." (2)

AN ANCIENT AFRICAN COIN IN AUSTRALIA

In June 2013 a Fortean news item broke in the mainstream media for a change, concerning a coin from eastern Africa found on an island near Australia's Northern Territory. Here is how the CNN news web-site reported the story on June 27th,(abridged version).

(CNN) - Can a handful of an ancient copper coins from a once-opulent but now abandoned corner of Africa change what we know about Australian history?

A team of researchers is on a mission to find out. With its glimmering wealth, busy harbour and coral stone buildings, the island of Kitwa hosted traders from as far away as China, who would exchange gold,ivory and iron from southern Africa's interior for Arabian pottery and Indian textiles as well as perfumes, porcelains and spices from the Far East…But interest in this nearly forgotten East African city has resurfaced lately thanks to the mystery surrounding a remarkable discovery thousands of miles away, in a long-abandoned , remote chain of small islands near Australia's Northern Territory.

Astonishing Discovery

Back in 1944, an Australian soldier named Maurie Isenberg was assigned
to one of the uninhabited but strategically positioned Wessel Islands to
man a radar station. One day, whilst fishing on the beach during his spare
time, he discovered nine coins buried in the sand. Isenberg stored them in
a tin until 1979, when he wondered if they might be worth something and
and sent them to be identified.

Four of the coins were found to belong to the Dutch East India Company,
with one of them being from the late 17th century... "It's a very
fascinating discovery" says Ian McIntosh, an Indiana University-Purdue
University Indianapolis anthropologist.

"Kilwa coins have only ever been found outside of the Kilwa region on
one or two occasions" he explains.

"A single coin was found in the ruins of great Zimbabwe and one coin
was found in the Arabian Peninsula, in what is now Oman, but nowhere
else. And yet, here is now this handful of them in northern Australia, this
is the astonishing thing" (3)

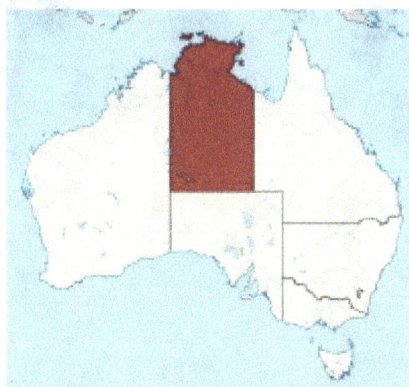

Fig 1 Australia's Northern
Territory Wikipedia Creative
Commons

True and Wonderfull.

A Difcourfe relating a ftrange and mon-
ftrous Serpent (or Dragon) lately difcouered, and yet
liuing, to the great annoyance and diuers flaughters
both of Men and Cattell, by his ftrong
and violent poyfon,

In Suffex *two miles from* Horfam, *in a woode*
called S. Leonards Forreft, and thirtie miles from
London, *this prefent month of Auguft.* 1614.
With the true Generation of Serpents.

Printed at London by *Iohn Trundle,*

Fig 2 Front page of pamphlet depicting Horsham Serpent.
It looks crocodile-like to me. Thanks to Bob Skinner for
providing this.

A BEAR-DOG HYBRID?

The following article appeared in the **Weekly Reno Gazette** of November 22nd 1883:

A QUEER CASE

EVIL ASSOCIATIONS CORRUPT GOOD MORALS[1]

On a ranch near Petaluma is an animal that is unmistakably a cross between a California brown bear[2] or a cinnamon bear and a dog. It was caught when young in the mountains near Salinus, fourteen years ago, and has been kept by its present owners ever since. The *Courier* says of it : In colour and appearance, when laying down, it would be taken for a bear, but when standing the dog cross is plainly visible. The head and tail show the bruin and feet are about half bear and half dog, except the crook of the hind-leg, which is longer and more dog-like. His habits are more of the bear order, being morose and sulky, and during the Winter inclined to sleep, and is ugly when disturbed. His bark is that of a Newfoundland dog, while his growl is that of a bear. Altogether it is a singular beast and an animal cross we never supposed existed. (4)

 plus  = ?

Fig 3 and *Fig 4*

Brown bear and Newfoundland dog.Images from Wikipedia Creative

---

[1] This is the real headline,not me moralising!

[2] The last Californian Brown Bear became extinct in 1922

Here is the exchange of messages on Facebook about this supposed bear-dog hybrid:

Richard Muirhead (May 5th 2013) : Has a cross between a bear and a Newfoundland dog ever been reported?

Martin Kilmer-True: Wat

Kirst Mason D`Raven: You mean something that looks like a cross between a Newfoundland and a bear?

Seamus Ehrhard: Yes, the great New-Bear has been sighted but never officially documented.

Matija Bobi: Bear and Newfoundland genes just wont splice.

Richard Muirhead: Hi Kirst yes that`s right (without giving the whole say (I meant story) away as its coming out in no. 5 of my magazine Flying Snake) I found a story in an old U.S paper c. 1891 of a cross between the now extinct Californian brown bear and a Newfoundland. Please can you tell me more about the great New - Bear Seamus?

Richard Muirhead: whole STORY I meant!

Tim Bergmann: I`m certain that`s genetically impossible.

Seamus Ehrhard: I`m actually not allowed to comment on the matter since a paper is now in review

Richard Muirhead: I am sure you are right Tim, I`m not saying it actually did happen, just that it was reported to have happened.

Richard Muirhead: OK Thanks Seamus.

Tim Bergmann: Richard I see…but there have always been such funny reports all around the world(not only at April 1ˢᵗ)…it`s just impossible…that`s all…probably a hoax.

David Smith: I`ve seen the photo pretty sure its photo shopped

Kirst Mason D`Raven: Is this stemming from the folklore that the Newfoundland dog breed originally came from the Viking `Bear Dogs`?

Johan Schmooley Moufasa: Absolutely impossible genetically. Pure fallacy.

Richard Muirhead: I don't recall seeing a photo, this was in 1891 remember. You must be thinking of something else David. (5)

HARE AT SEA

The Countryman Spring 1982 page 136

Some friends, returning from a November sea-fishing trip east of Worthing, went to investigate a small animal swimming about 100 yards offshore. When they got it aboard, they were surprised to find it was a hare. Far from being grateful for its rescue, it dashed wildly around the boat until it managed to escape overboard. They continued towards the shore and the hare followed them in. When another group of fishermen approached the hare, as it was recovering in the shelter of a groyne, it speedily made off up the pebble beach,ran several hundred yards along the main road, then turned up a side-road towards the downs. During the whole episode the hare did not appear to be in any distress except when approached or interfered with. We wondered whether it could have been carried out to sea through one of the drainage pipes - Mr & Mrs K.E. Weller, Sussex. [Winwood Reade, who has been studying the brown hare for some years, tells me she has not heard of a parallel occurrence, although hares are known to be good swimmers in fresh water and the early nineteenth-century naturalist Yarrell saw one cross `an arm of the sea a mile broad`. -B.C] (6)

A BLACK CAT IN QUEENSLAND
STRANGE ANIMAL SHOT

The Brisbane Courier September 17th 1924

Some days ago, writes our Tenterfield correspondent,Maurice Clark, the son of Mr A.Clark,of Steinbrook, shot a strange animal near his home,which has caused considerable speculation amongst local residents. The boy was out shooting, when he noticed something move near a log. He fired a bullet into animal, which immediately turned on him, and assumed such a ferocious aspect that he quickly returned home and informed his father. Mr Clark went out and killed the animal, which was of the cat species,and about the size of a full-grown fox. It was pure black in colour,and covered with a long fur,besides having a long thin tail. No local bushmen have ever seen anything of the kind before. What is believed to be the mate of the animal, which has been destroyed, can be heard at night in the vicinity of Steinbrook howling weirdly. Residents in the vicinity are somewhat afraid of it, and many shooting parties have been organised in the hope of destroying it, but without success so far. Some members of Wirth's Circus have expressed the opinion that possibly the animal was a young puma. (7)

GIANT SPECIES OF PYTHON
REPORTED

The Advertiser, Adelaide ,March 31st 1953

Giant Species Of Python Reported

DARWIN, March 30. Details were given here to-day by natives from the Djimba area of Central Arnhem Land of a giant species of rock python which grows to a length of 30 ft. and 18 in. across the body.

The natives claim the head of the snake is too big to fit into a kerosene tin.

One killed recently had four undigested wallabies inside.

A number of natives confirmed the size of the snake by stepping out 10 paces when questioned separately.

Fig 5 Image of a huge shark, somewhere
in Asia, from a Facebook site. I tried to find
out who owned the copyright to this but

Three items from **The Metro** June 5th 2013:

ZEBRA FINCHES, ETC:

GENE-IUS: Were you born with your personality or are
you parents responsible for it? If zebra finches are to be
relied on, its nurture (not nature) that counts. Exeter
university researchers found that counts. Exeter university
researchers found that finch foster parents exert more
influence on behaviour than the birds` genes.

DEATH DUTY: Dying doesn`t stop male Trinidadian
guppies having children.But it`s the females who do the
hard work.They can store sperm from their recently dec

-eased partners in their reproductive tracts. When they want to conceive,it's ready on tap, the Royal Society reports.

EARLY BIRDS: City living has changed our day and time activities…and the same can be said for songbirds. They rise earlier than their country cousins and have faster `body clocks`. Urban life has changed their biological rhythms, a German study published by the Royal Society suggests. (8)

A TIGER IN HONG KONG IN 1940

 I received the following account via e-mail from Brian Edgar of the Yahoo Stanley P.O.W. Camp group , Hong Kong, 1941-45 .It is the account of someone called Sheridan's visit to Lantau in 1940,two years before the famous (or should that be infamous?) case of the tigers who visited the camp in May 1942.

James and I were dressed in KD shorts and shirt, walking shoes and I had a small white sun hat. He carried a water bottle and small haversack as well as a camera and binoculars. Towards evening we returned to the Monastery rather leg weary, but having enjoyed our first day on a delightful island. Our monk friend greeted us with two tubs of hot water, in which we had a good wash down and soaked our feet in. After cooking our evening meal we had a rest, and then decided to take our sleeping kit and sleep out on the hill above the Monastery. As we were about to leave, our monk friend came with the Head Monk. Both seemed very agitated and tried to persuade us to stay inside the Monastery compound. In the course of their conservation which was all in Chinese , I could understand the word which meant tiger and by further signs they indicated that there were tigers on the island and that it was dangerous for us to sleep in the open . We of course had never heard of tigers in any part of South China, [1] so we just laughed and prepared to leave. The head Monk then said " man-man" i.e. wait a minute, then fetched two staves about 5ft. long with a wooden shaft and steel point. We took them to please him and made our way up the tracks to where we had decided to sleep. We noticed the gates were closed as we left. It was very pleasant, as darkness set in, we could see the lights of Macau and of ships passing in the distance. James and I slept sound,

---

[1] This may seem dubious but actually this opinion was surprisingly common in Hong Kong at the time.

although it felt a bit hard underneath. We woke to a fine clear cool morning, picked up our gear and made our way back to the Monastery. They must have had some sort of look-out because we could see a group at the gate as we approached. It was the head Monk who greeted us, and he seemed very agitated. He motioned us to follow him to a sort of shrine which was about 4400 yards outside the Monastery. At the shrine on the ground lay a large Chow dog of which quite a few were kept in the Monastery. It was dead and torn to pieces, the head Monk pointed to it and said one word in Chinese, "tiger". James and I looked at each other, and said, "no more sleeping in the open." (9)

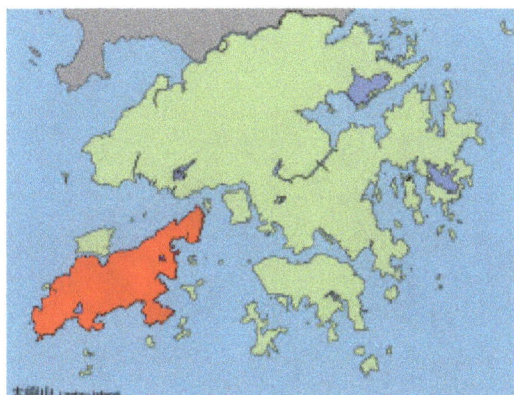

Fig 6 Wikipedia Creative Commons.

Lantau is the large island in the south-
west of the picture.

REFERENCES

1. Knowledge Magazine   October 26th 1883

2. The Guardian March 14th 2013

3. See Ancient coins that could change history of Australia
edition.cnn.com/2013/06/25/world/africa/ancient-african-coins-

4 Weekly Reno Gazette November 22nd 1883

5. Facebook communicatons May 5th 2013

6. The Countryman Spring 1982 p. 136

7. The Brisbane Courier September 17th 1924

8. The Metro June 5th 2013

9. Yahoo Stanley Camp Forum e-mail from Brian Edgar May 9th 2013

PHOTO OF A COELACANTH TAKEN BY BRITTA RODDE, COMOROS ISLANDS, 1995

Fig 7 © Britta Rodde.

This photo of a coelacanth appeared on Britta Rodde`s Facebook wall and I was kindly given permission to reproduce by her. It was taken in the Comoros Islands in 1995.

A NEW ZEALAND SEA MONSTER, RECENTLY CAUGHT AT THE THAMES, AUCKLAND

This curious fish, which has some of the characteristics of the seal and the shark, is evidently a voracious creature. In its stomach were found ten snapper, one yellowtail, and a fourfoot dog fish.   From a sketch by Harry L. Wright.

Fig 8 The caption under the image, dated 31/10/1901, reads: `A New Zealand Sea Monster Recently Caught At The Thames, Auckland. This curious fish,which has some of the characteristics of the seal and the shark,is evidently a voracious creature. In its stomach were found ten snapper, one yellow tail, and a four foot dog-fish.` On Facebook, Britta Rodde believed it was a basking shark,Paul Fitzpatrick a Greenland Shark

Reproduced with permission of Sir George Grey Special Collections Auckland Libraries. AWNS 19011031-7-2

70

EXAMPLES OF SOME OF THE STRANGE INSECTS WHICH FELL IN BATH IN 1871 (FROM SCIENCE GOSSIP)

HARDWICKE'S SCIENCE-GOSSIP. 229

Fig 9

INSECTS AT BATH.

A CORRESPONDENT inquires concerning the nature of the insects which fell lately at Bath. I beg to inclose a photograph of a drawing of them, made by a friend of mine in Bath, a gentleman well versed in natural history.

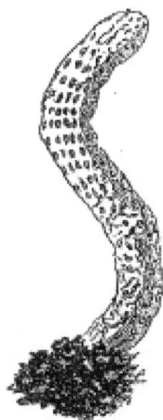

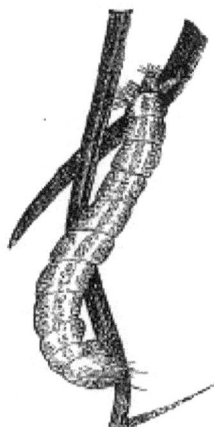

Fig. 138.

Fig. 139.

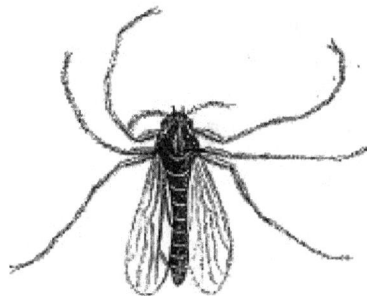

Fig. 140.

Copy from Photograph of insects that fell at Bath.

W. B. GIBBS.

More information about these insects can be found in Charles Fort's 'Book of The Damned'

Letters to Flying Snake

GREY SQUIRRELS - A NEW DISEASE?

May 20th 2013

An interesting contribution from regular Richard George of St Albans on a possible new disease amongst grey squirrels:

Dear Richard

"I've been reading a book by Neil Ansell called *Deep Country* ,(Hamish Hamilton,2011), which describes his five years living in an isolated cottage in north Breconshire near the town of Sennybridge. His observation of the natural world around him is both acute and poetic. On pages 70-1 there is a description of an illness in a squirrel which you might find interesting:

I was sitting on my doorstep in the morning sun with my chain file, methodically sharpening my saw, when I noticed something crawling up the hillside towards me. It was a grey squirrel, but its head was swollen,its eyes bulging, and it appeared to be losing control of its limbs. I hadn't realized that myxomatosis could affect squirrels; or perhaps this was not myxomatosis but an equivalent squirrel disease. It couldn't support its weight; it was dragging itself up the hill on its belly, but it seemed absolutely determined and it never paused in its struggle…

The squirrel carried on,painfully slowly,irrevocably. Just over the track was an old ash,and there was a hollow in that ash right at ground level. The squirrel crawled into the shelter of that hole, from which it would never emerge. "

A SHOWER OF PERIWINKLES IN WORCESTER
IN 1881

On August 17th 2010 I received the following letter from Mike Rowe, Lymington after I began looking into a shower of periwinkles at Worcester on May 28th 1881. I was going to write up my research for the Internatonal Journal of Meteorolgy but I never did.

Dear Richard

"Many thanks for your letter about the shower of periwinkles at Worcester on 28 May 1881. This fall is at least as peculiar as the 1867 shower of nuts in Ireland. (I liked your article on this in *IJ Met* by the way.) There is nothing about the Worcester case in *British Rainfall* for 1881, and there doesn't seem to be anything in *Symons's Monthly Meteorological Magazine for 1881* either. It's clear from British Rainfall that it was a thundery day, with some severe storms in the west Midlands.

The only primary sources I can find is a report from the Worcester Daily Times, 30 May 1881, reproduced in *Phenomena: A Book of Wonders*, by John Michell and Robert J.M. Rickard, Thames and Hudson, 1977. I enclose a photocopy of the whole section 'Falls of creatures and organic matter' so that you can judge the general reliability of the book!

I think the case also features in one or more of William Corliss's books.

If these moluscs actually were periwinkles , then either they came from the sea,by the agency of a tornado, or they were discarded by a local fisherman and the story has been exaggerated. Periwinkles certainly were (and are) eaten. The tornado theory seems impossible to me because of the distance involved. I did wonder,years ago, whether the winkles could have been the land snail *Pomatias elegans,* which looks like a winkle and - unlike almost all land snails - even has an operculum or door with which to close the mouth of the shell, as a winkle has. (I was very interested in land and freshwater molluscs when I was in the Sixth Form at school, and sent a lot of records to a mapping scheme being conducted by the Conchological Society.) But *P.elegans* is a local species largely confined to limestone, and I doubt whether it's found within some miles of Worcester. The nearest known location is the Bredon Hills area, c 10 km from Worcester. In any case, it wouldn't explain the hermit crab mentioned in the press article - although this does sound to me like an embellishment to the story! I'm sure I've read somewhere that winkles were found on roofs and walls, but again, how reliable is this information?

On the Web, there is some material in the hypertext edition of Charles Fort's *Lo!*

I hope this is of interest, and also that you can find other primary sources on this event. Its'fascinating!

Best wishes

Mike

Fig 1 Pomatias elegans. Wikipedia Creative Commons

BOOK REVIEW

Footwear Markings at Bolsover Castle, Derbyshire . A Report for English Heritage. Richard Sheppard Nottingham: Trent and Peak Archaeological Trust. 1997

This utterly vital,groundbreaking and bold report on the increasingly important subject of footwear markings landed on my door mat in September 1997 in Oxford. Richard Sheppard is or was a bit of a Fortean himself .He wrote to me, on the 12th of that month: " Dear Sir, Please find enclosed a copy of a report that I sent to English Heritage earlier this year. Whilst researching the subject of footwear markings on lead roofs, I did send a letter to `Letters to Ambrose Merton` and yours is the first response that I have had - albeit asking for information, rather than offering any! Anyway,what I have written only touches on this interesting subject and here in the context of Bolsover. I have submitted a shortened and revised paper to Derbyshire Archaeological Journal for publication next year. Please note that copyright of this report`s contents lies with us (Trent & Peak Archaeological Trust) and English Heritage.

As regards black squirrels, I`m afraid that I cannot help you there. I`m told that they exist in Russia or the Ukraine, but otherwise I know nothing about them being seen in the U.K. One for the `Fortean Times` I suspect!" Well it is a fascinating document, 25 pages of charts and images ,mainly of outlines of footwear, some with letters on them,others names and dates. The dates range from "at least 1822 to the early 20th century." "By far the majority of the samples were left by men. Of the complete 140 outlines, about 20 were probably left by women, with a similar number being probable youths` shoes. Being of comparable lengths these latter two groups can be difficult to distinguish, although waists and heels are useful pointers…Leaving one`s mark on a lead roof may not have been looked upon as wanton vandalism at this time, as it might today. Inscribing graffiti on deserted old buildings was quite common in the 19th century, but Bolsover Castle has little evidence of defacement on its actual walls…This leaves the intriguing question: was leaving one`s individual mark on the Bolsover roof seen as an innocent activity or was it an act of bedevilment, a way of `cocking a snoop` at a snobbish vicar under his very nose." Fascinating stuff and he`s a Richard, to boot! (Groan!). I wonder if it went to a 2nd edition?

To be a Naturalist is better then to be a King

www.steampunknaturalist.com

One of a collection of odd coloured moles found in Berkshire in 2012 and passed on to Carl Marshall. See Animals & Men 50 for more info.

STILL ON THE TRACK OF UNKNOWN ANIMALS

T he Centre for Fortean Zoology, or CFZ, is a non profit-making organisation
founded in 1992 with the aim of being a clearing house for information, and coordi-
nating research into mystery animals around the world.

We also study out of place animals, rare and aberrant animal behaviour, and Zooform Phenomena; little-understood "things" that appear to be animals, but which are in fact nothing of
the sort, and not even alive (at least in the way we understand the term).

Not only are we the biggest organisation of our type in the world, but - or so we like to think
- we are the best. We are certainly the only truly global cryptozoological research organisation, and we carry out our investigations using a strictly scientific set of guidelines. We are
expanding all the time and looking to recruit new members to help us in our research into
mysterious animals and strange creatures across the globe.

Why should you join us? Because, if you are genuinely interested in trying to solve the last
great mysteries of Mother Nature, there is nobody better than us with whom to do it.

Members get a four-issue subscription to our journal *Animals & Men.* Each issue contains nearly 100 pages packed with news, articles, letters, research papers, field reports, and even a gossip column! The magazine is Royal Octavo in format with a full colour cover. You also have access to one of the world's largest collections of resource material dealing with cryptozoology and allied disciplines, and people from the CFZ membership regularly take part in fieldwork and expeditions around the world.

The CFZ is managed by a three-man board of trustees, with a non-profit making trust registered with HM Government Stamp Office. The board of trustees is supported by a Permanent Directorate of full and part-time staff, and advised by a Consultancy Board of specialists - many of whom are world-renowned experts in

their particular field. We have regional representatives across the UK, the USA, and many other parts of the world, and are affiliated with other organisations whose aims and protocols mirror our own.

You'll find that the people at the CFZ are friendly and approachable. We have a thriving forum on the website which is the hub of an ever-growing electronic community. You will soon find your feet. Many members of the CFZ Permanent Directorate started off as ordinary members, and now work full-time chasing monsters around the world.

Write to us, e-mail us, or telephone us. The list of future projects on the website is not exhaustive. If you have a good idea for an investigation, please tell us. We may well be able to help.

We are always looking for volunteers to join us. If you see a project that interests you, do not hesitate to get in touch with us. Under certain circumstances we can help provide funding for your trip. If you look on the future projects section of the website, you can see some of the projects that we have pencilled in for the next few years.

In 2003 and 2004 we sent three-man expeditions to Sumatra looking for Orang-Pendek - a semi-legendary bipedal ape. The same three went to Mongolia in 2005. All three members started off merely subscribers to the CFZ magazine. Next time it could be you!

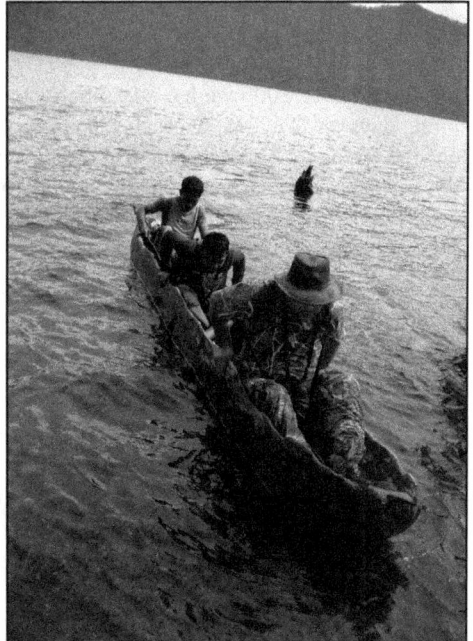

We have no magic sources of income. All our funds come from donations, membership fees, and sales of our publications and merchandise. We are always looking for corporate sponsorship, and other sources of revenue. If you have any ideas for fund-raising please let us know. However, unlike other cryptozoological organisations in the past, we do not live in an intellectual ivory tower. We are not afraid to get our hands dirty, and furthermore we are not one of those organisations where the membership have to raise money so that a privileged few can go on expensive foreign trips. Our research teams, both in the UK and abroad, consist of a mixture of experienced and inexperienced personnel. We are truly a community, and work on the premise that the benefits of CFZ membership are open to all.

Reports of our investigations are published on our website as soon as they are available. Preliminary reports are posted within days of the project finishing.

Each year we publish a 200 page yearbook containing research papers and expedition reports too long to be printed in the journal. We freely circulate our information to anybody who asks for it. We have a thriving YouTube channel, CFZtv, which has well over two hundred self-made documentaries, lecture appearances, and episodes of our monthly webTV show. We have a daily online magazine, which has over a million hits each year.

Each year since 2000 we have held our annual convention - the Weird Weekend. It is three days of lectures, workshops, and excursions. But most importantly it is a chance for members of the CFZ to meet each other, and to talk with the members of the permanent directorate in a relaxed and informal setting and preferably with a pint of beer in one hand. Since 2006 - the Weird Weekend has been bigger and better and held on the third weekend in August in the idyllic rural location of Woolsery in North Devon.

Since relocating to North Devon in 2005 we have become ever more closely involved with other community organisations, and we hope that this trend will continue.

We have also worked closely with Police Forces across the UK as consultants for animal mutilation cases, and we intend to forge closer links with the coastguard and other community services. We want to work closely with those who regularly travel into the Bristol Channel, so that if the recent trend of exotic animal visitors to our coastal waters continues, we can be out there as soon as possible.

Apart from having been the only Fortean Zoological organisation in the world to have consistently published material on all aspects of the subject for over a decade, we have achieved the following concrete results:

- Disproved the myth relating to the headless so-called sea-serpent carcass of Durgan beach in Cornwall 1975
- Disproved the story of the 1988 puma skull of Lustleigh Cleave
- Carried out the only in-depth research ever into the mythos of the Cornish Owlman.
- Made the first records of a tropical species of lamprey
- Made the first records of a luminous cave gnat

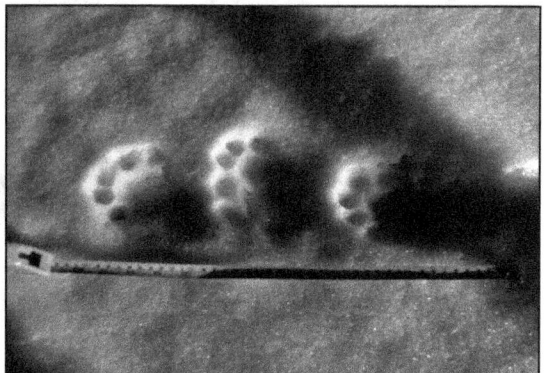

larvae in Thailand
- Discovered a possible new species of British mammal - the beech marten
- In 1994-6 carried out the first archival fortean zoological survey of Hong Kong
- In the year 2000, CFZ theories were confirmed when a new species of lizard was added to the British List
- Identified the monster of Martin Mere in Lancashire as a giant wels catfish
- Expanded the known range of Armitage's skink in the Gambia by 80%
- Obtained photographic evidence of the remains of Europe's largest known pike
- Carried out the first ever in-depth study of the ninki-nanka
- Carried out the first attempt to breed Puerto Rican cave snails in captivity
- Were the first European explorers to visit the `lost valley` in Sumatra
- Published the first ever evidence for a new tribe of pygmies in Guyana
- Published the first evidence for a new species of caiman in Guyana

- Filmed unknown creatures on a monster-haunted lake in Ireland for the first time
- Had a sighting of orang pendek in Sumatra in 2009
- Found leopard hair, subsequently identified by DNA analysis, from rural North Devon in 2010
- Brought back hairs which appear to be from an unknown primate in Sumatra
- Published some of the best evidence ever for the almasty in southern Russia

CFZ Expeditions and Investigations include:

- 1998 Puerto Rico, Florida, Mexico (Chupacabras)

- 1999 Nevada (Bigfoot)
- 2000 Thailand (Naga)
- 2002 Martin Mere (Giant catfish)
- 2002 Cleveland (Wallaby mutilation)
- 2003 Bolam Lake (BHM Reports)
- 2003 Sumatra (Orang Pendek)
- 2003 Texas (Bigfoot; giant snapping turtles)
- 2004 Sumatra (Orang Pendek; cigau, a sabre-toothed cat)
- 2004 Illinois (Black panthers; cicada swarm)
- 2004 Texas (Mystery blue dog)
- Loch Morar (Monster)
- 2004 Puerto Rico (Chupacabras; carnivorous cave snails)
- 2005 Belize (Affiliate expedition for hairy dwarfs)
- 2005 Loch Ness (Monster)
- 2005 Mongolia (Allghoi Khorkhoi aka Mongolian death worm)
- 2006 Gambia (Gambo - Gambian sea monster , Ninki Nanka and Armitage's skink
- 2006 Llangorse Lake (Giant pike, giant eels)
- 2006 Windermere (Giant eels)
- 2007 Coniston Water (Giant eels)
- 2007 Guyana (Giant anaconda, didi, water tiger)
- 2008 Russia (Almasty)
- 2009 Sumatra (Orang pendek)
- 2009 Republic of Ireland (Lake Monster)
- 2010 Texas (Blue Dogs)
- 2010 India (Mande Burung)
- 2011 Sumatra (Orang-pendek)
- 2012 Sumatra (Orang Pendek)
- 2014 Tasmania (Thylacine)
- 2015 Tasmania (Thylacine)
- 2016 Tasmania (Thylacine)
- 2017 Tasmania (Thylacine)

For details of current membership fees, current expeditions and investigations, and voluntary posts within the CFZ that need your help, please do not hesitate to contact us.

The Centre for Fortean Zoology,
Myrtle Cottage,
Woolfardisworthy,
Bideford, North Devon
EX39 5QR

Telephone 01237 431413
Fax+44 (0)7006-074-925
eMail info@cfz.org.uk

Websites:

www.cfz.org.uk
www.weirdweekend.org

THE WORLD'S

Edited by Jon Downes

WEIRDEST

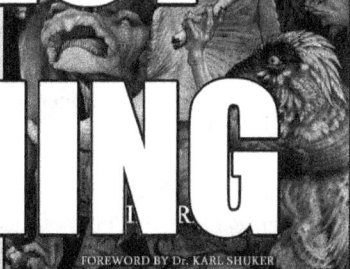

Jonathan Downes and Richard Freeman

FOREWORD BY Dr. KARL SHUKER

PUBLISHING

COMPANY

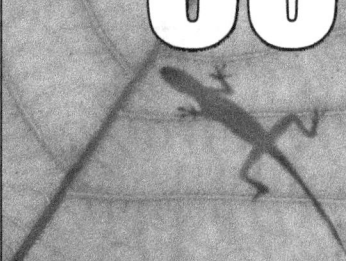

CARL PORTMAN

THE COLLECTED POEMS
Dr Karl P. N. Shuker

an anthology of writings by
ANDY ROBERTS

HOW TO START A PUBLISHING EMPIRE

Unlike most mainstream publishers, we have a non-commercial remit, and our mission statement claims that "we publish books because they deserve to be published, not because we think that we can make money out of them". Our motto is the Latin Tag *Pro bona causa facimus* (we do it for good reason), a slogan taken from a children's book *The Case of the Silver Egg* by the late Desmond Skirrow.

WIKIPEDIA: "The first book published was in 1988. *Take this Brother may it Serve you Well* was a guide to *Beatles* bootlegs by Jonathan Downes. It sold quite well, but was hampered by very poor production values, being photocopied, and held together by a plastic clip binder. In 1988 A5 clip binders were hard to get hold of, so the publishers took A4 binders and cut them in half with a hacksaw. It now reaches surprisingly high prices second hand.

The production quality improved slightly over the years, and after 1999 all the books produced were ringbound with laminated colour covers. In 2004, however, they signed an agreement with Lightning Source, and all books are now produced perfect bound, with full colour covers."

Until 2010 all our books, the majority of which are/were on the subject of mystery animals and allied disciplines, were published by `CFZ Press`, the publishing arm of the Centre for Fortean Zoology (CFZ), and we urged our readers and followers to draw a discreet veil over the books that we published that were completely off topic to the CFZ.

However, in 2010 we decided that enough was enough and launched a second imprint, `Fortean Words` which aims to cover a wide range of non animal-related esoteric subjects. Other imprints will be launched as and when we feel like it, however the basic ethos of the company remains the same: Our job is to publish books and magazines that we feel are worth publishing, whether or not they are going to sell. Money is, after all - as my dear old Mama once told me - a rather vulgar subject, and she would be rolling in her grave if she thought that her eldest son was somehow in `trade`.

Luckily, so far our tastes have turned out not to be that rarified after all, and we have sold far more books than anyone ever thought that we would, so there is a moral in there somewhere...

Jon Downes,
Woolsery, North Devon
July 2010

CFZ PRESS

CFZ Press is our flagship imprint, featuring a wide range of intelligently written and lavishly illustrated books on cryptozoology and the quirkier aspects of Natural History.

CFZ Classics is a new venture for us. There are many seminal works that are either unavailable today, or not available with the production values which we would like to see. So, following the old adage that if you want to get something done do it yourself, this is exactly what we have done.

Desiderius Erasmus Roterodamus (b. October 18th 1466, d. July 2nd 1536) said: "When I have a little money, I buy books; and if I have any left, I buy food and clothes," and we are much the same. Only, we are in the lucky position of being able to share our books with the wider world. CFZ Classics is a conduit through which we cannot just re-issue titles which we feel still have much to offer the cryptozoological and Fortean research communities of the 21st Century, but we are adding footnotes, supplementary essays, and other material where we deem it appropriate.

http://www.cfzpublishing.co.uk/

Fortean Words is a new venture for us. The F in CFZ stands for "Fortean", after the pioneering researcher into anomalous phenomena, Charles Fort. Our Fortean Words imprint covers a whole spectrum of arcane subjects from UFOs and the paranormal to folklore and urban legends. Our authors include such Fortean luminaries as Nick Redfern, Andy Roberts, and Paul Screeton. . New authors tackling new subjects will always be encouraged, and we hope that our books will continue to be as ground-breaking and popular as ever.

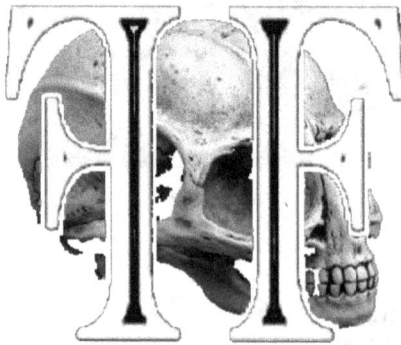

Just before Christmas 2011, we launched our third imprint, this time dedicated to - let's see if you guessed it from the title - fictional books with a Fortean or cryptozo-ological theme. We have published a few fictional books in the past, but now think that because of our rising reputation as publishers of quality Forteana, that a dedicated fiction imprint was the order of the day.

http://www.cfzpublishing.co.uk/

www.ingramcontent.com/pod-product-compliance
Lightning Source LLC
Chambersburg PA
CBHW050449270326
41927CB00009B/1670